AF358736

Design of Heat Exchangers for Heat Pump Applications

Design of Heat Exchangers for Heat Pump Applications

Editors

Marco Fossa
Antonella Priarone

MDPI • Basel • Beijing • Wuhan • Barcelona • Belgrade • Manchester • Tokyo • Cluj • Tianjin

Editors
Marco Fossa
University of Genova
Italy

Antonella Priarone
University of Genova
Italy

Editorial Office
MDPI
St. Alban-Anlage 66
4052 Basel, Switzerland

This is a reprint of articles from the Special Issue published online in the open access journal *Energies* (ISSN 1996-1073) (available at: https://www.mdpi.com/journal/energies/special_issues/ heat_exchangers_heat_pump).

For citation purposes, cite each article independently as indicated on the article page online and as indicated below:

LastName, A.A.; LastName, B.B.; LastName, C.C. Article Title. *Journal Name* **Year**, *Article Number*, Page Range.

ISBN 978-3-03943-513-5 (Pbk)
ISBN 978-3-03943-514-2 (PDF)

Contents

About the Editors

Marco Fossa, Ph.D., is full Professor at Dime Department (www.dime.unige.it) of the University of Genova, Italy. He is in charge of M.Sc. classes of Applied Thermodynamics and Heat Transfer, Renewable Energies, Solar and Geothermal Energy. He is the Coordinator of the M.Sc. course in Energy Engineering (www.en2.unige.it) since its constitution, year 2014. Research fields include: Geothermal heat pumps, Solar Energy, High Efficiency Greenhouses, Heat Transfer and Fluid Flow Measurements, Thermal Energy System Modelling. Visiting Appointments: The University of the New South Wales, Sydney (as Visiting Professor): 2006, 2008, 2010, 2012, 2017Cern, Geneva: 1991, 1992, 1997–2005 (CMS particle detector project, as Research Fellow) University of Nottingham, SChEME, 2001. MCI University Innsbruck (as Visiting Professor 2018, 2019, 2020). Other international collaborations (year 2020): Polytech Savoie, University of the New South Wales (Unsw, Sydney), KTH Stockholm, Polytech Montreal Member of international Ph.D. juries: Unsw Sydney (2011, 2012, 2019), Univ. Sydney (2015), Univ. Tetouan (2010), Univ. Lyon 1 (2007. 2011, 2015, 2016, 2017) Member of International Professor Position Juries: Univ. Lyon 1 (2014), University of the New South Wales, Sydney (2017, 2019). Prof. Fossa is author of about 150 scientific papers in the field of Thermal Engineering. Scopus citations = 1000, Scopus h-index = 17.

Antonella Priarone obtained her Master's Degree in Mechanical Engineering in 2001 at the University of Genova, and she received her Ph.D. in Applied Thermodynamics and Heat Transfer in 2005. She is Researcher-Assistant Professor of Applied Thermodynamics and Heat Transfer at the University of Genova since 1 July 2010. She is Professor of the courses "Heat Transfer" and "Energy and Buildings" for the Master Degree in Energy Engineering. She has faced several research subjects ranging from heat transfer fundamentals to advanced engineering applications; the approach to the study was alternately experimental and numerical. In detail, she was involved in experimental study on nucleate pool boiling and on two-phase pressure drop in minichannels. Currently, she is concerned with numerical models of coaxial and U-tube ground-coupled heat exchangers, of Thermal Response Tests and with numerical study of long-term performance of BHE fields, with or without groundwater movement. Moreover, she is interested also in buildings energy simulation with E-plus, with reference to the special case of agricultural greenhouses.

Preface to "Design of Heat Exchangers for Heat Pump Applications "

Heat pumps (HPs) allow for providing heat without direct combustion, in both civil and industrial applications. They are very efficient systems that, by exploiting electrical energy, greatly reduce local environmental pollution and CO_2 global emissions. The fact that electricity is a partially renewable resource and because the coefficient of performance (COP) can be as high as four or more, means that HPs can be nearly carbon neutral for a full sustainable future.

The proper selection of the heat source and the correct design of the heat exchangers is crucial for attaining high HP efficiencies—examples can be ground coupled heat exchangers, lake/sea/waste water systems, enhanced surface heat exchangers, and HPs exploiting waste heat from industrial and civil processes.

Heat exchangers (also in terms of HP control strategies) are hence one of the main elements of HPs, and improving their performance enhances the effectiveness of the whole system. Both the heat transfer and pressure drop have to be taken into account for the correct sizing, especially in the case of mini- and micro-geometries, for which traditional models and correlations can not be applied. New models and measurements are required for best HPs system design, including optimization strategies for energy exploitation, temperature control, and mechanical reliability. A relevant feature is also the phase change of the refrigerant, which can involve problems related to the phase distribution in the heat exchanger. Moreover, the selection of the proper refrigerant fluid it is important in order to improve the energy performance and to enhance environmental compatibility. Thus, a multidisciplinary approach of the analysis is requested.

Marco Fossa, Antonella Priarone
Editors

Article

An Enhanced Vertical Ground Heat Exchanger Model for Whole-Building Energy Simulation

Matt S. Mitchell [1,*] and Jeffrey D. Spitler [2]

[1]   National Renewable Energy Laboratory, Golden, CO 80401, USA
[2]   School of Mechanical and Aerospace Engineering, Oklahoma State University, Stillwater, OK 74078, USA; spitler@okstate.edu
*   Correspondence: matt.mitchell@nrel.gov

Received: 26 June 2020; Accepted: 23 July 2020; Published: 5 August 2020

Abstract: This paper presents an enhanced vertical ground heat exchanger (GHE) model for whole-building energy simulation (WBES). WBES programs generally have computational constraints that affect the development and implementation of component simulation sub-models. WBES programs require models that execute quickly and efficiently due to how the programs are utilized by design engineers. WBES programs also require models to be formulated so their performance can be determined from boundary conditions set by upstream components and environmental conditions. The GHE model developed during this work utilizes an existing response factor model and extends its capabilities to accurately and robustly simulate at timesteps that are shorter than the GHE transit time. This was accomplished by developing a simplified dynamic borehole model and then exercising that model to generate exiting fluid temperature response factors. This approach blends numerical and analytical modeling methods. The existing response factor models are then extended to incorporate the exiting fluid temperature response factor to provide a better estimate of the GHE exiting fluid temperature at short simulation timesteps.

Keywords: ground heat exchanger; whole-building energy simulation; ground source heat pump; g-Function

---

1. Introduction

Whole-building energy simulation (WBES) programs, such as EnergyPlus® [1] and TRNSYS [2], use equipment-loop simulation algorithms that pass component (e.g., pumps, boilers, chillers, heat exchangers, etc.) entering and exiting conditions (e.g., flow rates, temperatures, humidity ratios, etc.) flow-wise from up-stream components to down-stream components. These components are generally connected in a loop so that every component on the loop is in some way connected to every other component on the loop. Equipment that is active at any given timestep may be controlled by an individual controller or by a larger control algorithm that specifies control for the entire system. These components may also make requests to other components for specific performance. For example, a chiller model that has a minimum required flow rate may require that all other components on the loop run at its minimum flow rate, even though they could be operated at a lower flow rate.

Passive equipment, on the other hand, simply operates by providing what it can at the current given conditions. For a ground heat exchanger (GHE), the models are only expected to be passed entering conditions, which in this case is the circulating fluid's temperature and mass flow rate. As expected in the real world, the GHE will only give whatever performance it can physically deliver, regardless of what the other components on the loop require for successful operation. The objective of this study was to develop a GHE model that behaves in a similar passive manner, and to develop it for use in WBES environments. Requirements associated with this application are as follows.

First, the developed model must be accurate and produce physically realistic results. Some modeling simplifications may be necessary to meet secondary needs, such as simulation time. Regardless, the model should still produce a result that is as accurate as possible. In addition, WBES simulations often operate at non-uniform timesteps, so the model must accurately simulate GHE behavior at both short and long timesteps.

Second, the model should simulate GHE behavior as quickly and efficiently as possible. WBES simulations are often run on personal computers, such as laptops or desktop computers. These programs are generally not so sophisticated as to employ parallel- or multi-processing methods. Simulations are often run on a single computational process, which results in each sub-model within a WBES simulation being run in order, sequentially. Therefore, adding model complexity that significantly increases simulation time should only be done when absolutely necessary. Beyond this, modelers often use WBES to perform parametric analysis of designs, so any increase in simulation time has the potential to quickly amplify the total simulation time for any parametric study performed. This may be a challenge for designers who do not have access to a computer cluster.

Third, the model should be formulated to operate within a WBES environment. This means that the model should be formulated so the boundary conditions for the model can be clearly defined by the flow-wise upstream component and the external environment. For GHE, this implies that the GHE outlet fluid temperature can be computed directly from the inlet fluid's temperature and mass flow rate as well as by other user-defined or environmental parameters. GHE thermal loads are not expected to be known before hand.

Fourth, the model should not rely on libraries, software, or other resources that are not freely available or that must exist external to the simulation environment. Historical WBES GHE models have relied on third-party libraries or software tools to provide input data for the simulation. To avoid this additional burden on the modeler, the model developed here must be able to operate in a standalone manner.

2. Literature Review

A number of different GHE modeling methodologies have been developed over the years. As is typical for methods and technologies that evolve with time, these models have evolved from simple, closed-form analytical expressions to methods that are mathematically complex and simulate specific, nuanced physical behavior. Several reviews of these models are presented in [3,4], but we review them again here in the context of WBES for completeness.

2.1. Analytical Models

Analytical models form the backbone of GHE simulation and validation and have been used extensively to model systems containing individual boreholes. Analytical models also provide an "exact" solution, which can be used to validate more complicated models.

Historically, GHE modeling methodologies have treated the region inside the borehole as separate from the region outside the borehole. Ignoring for a moment the U-tube, grout, and circulating fluid, the GHE may be thought of as a line or cylinder conducting heat into the ground, which is assumed to be a semi-infinite medium with a uniform, constant temperature and isotropic properties.

The infinite line-source (ILS) model is a pure conduction solution for an infinite line that conducts heat continuously at a constant rate into an infinite medium. Some authors have discussed who originally developed the method and the limits of its applicability [5–8]. Mogensen [9] applied several approximations to the method to compute the mean borehole fluid temperature. Spitler & Gehlin [6] provide a form of the method that is commonly used for thermal response test (TRT) analysis. The ILS form provided by Spitler & Gehlin computes the mean fluid temperature response results for a single borehole with constant heat input rate.

Infinite cylinder source (ICS) analytical models are similar to ILS models; however, the geometry is treated as a cylinder where varying conditions can be applied inside the cylinder. Three different forms have been given by Carslaw & Jaeger [7]. Solution methods for ICS solutions can be computationally expensive and some authors have provided methods to reduce this burden [10,11]. Others have developed ICS and ILS advancements for modeling ground water flow, ice formation, and helical geometries [11–14].

Finite line-source (FLS) models are another analytical solution method. Like ILS and ICS models, FLS models are relatively easy to compute but allow for the ground surface and GHE end effects to be taken into account. FLS models function by integrating point-source solutions over finite line segments. The individual temperature response of each segment is calculated based on its interactions with all other segments, and the temperature response of the GHE is determined by taking the average of the temperature responses of each segment in the GHE.

FLS models have been developed to model arbitrary GHE configurations, [15]; however, these can be computationally expensive, so simplified forms have also been developed [16,17]. FLS models have been validated against a 3D finite element model and integrated into GHE design tools [15,18,19].

While the analytical ILS and ICS methods described previously are useful for TRT analysis and for model validation purposes, they are not generally suitable for GHE simulation in the context of WBES. The methods assume an infinite ground medium, so end effects at the ground surface and bottom of the boreholes are not captured. These become more important over time, particularly in predicting long-term heat build-up or draw-down. The models also require temporal superposition to handle time-varying heat inputs, and even then are formulated to require heat input rather than entering fluid temperature inputs. ILS and ICS models also have limited accuracy at short times. FLS solutions are commonly used for computing response factors, which, as discussed in the next section, are useful for WBES.

2.2. Response Factor Models

Response factor models—commonly referred to as "g-function" models—are another commonly used model for simulating GHE performance. The so called "g-function" curves represent the non-dimensional temperature rise of the borehole vs. non-dimensional logarithmic time as seen in Figure 1.

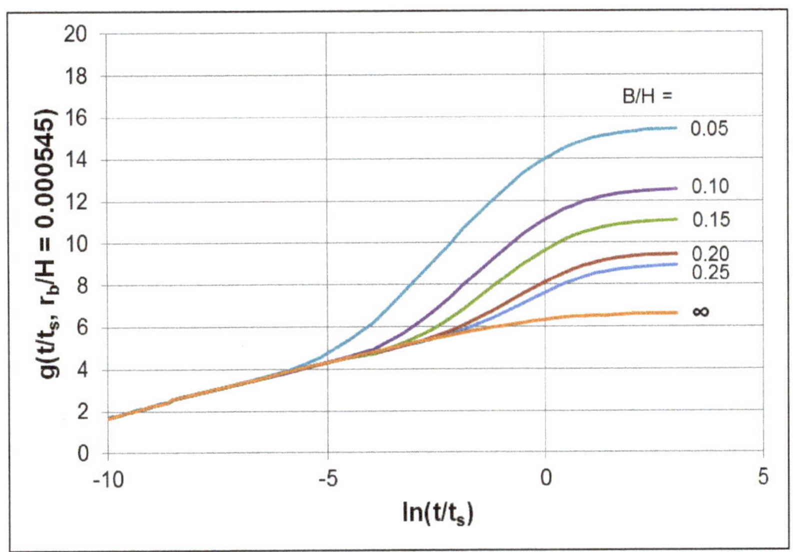

Figure 1. Example g-functions for a 3 × 2 rectangular GHE array.

These models apply Duhamel's theorem [20,21], which posits that the time-varying temperature solutions to inhomogeneous, partial differential conduction equations can be solved in a piecewise fashion by applying step-pulse inputs. In other words, GHE temperature response due to time-varying heat inputs can be determined by applying the superposition of piecewise heat constant heat pulses. Claesson [22] began applying the principle of superposition to GHE modeling, and later, along with his Ph.D. student Eskilson [23–25], developed the response factor models commonly used today. The average borehole wall temperature can be computed for a single, constant heat pulse from Equation (1) or for a series of piecewise heat pulses as shown in Equation (2):

$$T_b = T_s + \frac{q}{2\pi k_s^*} \cdot g\left(t/t_s, r_b/H, B/H, D/H\right) \tag{1}$$

$$T_b = T_s + \sum_{i=1}^{n} \frac{q_i - q_{i-1}}{2\pi k_s^*} \cdot g\left(\frac{t_n - t_{i-1}}{t_s}, r_b/H, B/H, D/H\right) \tag{2}$$

The g-functions are non-dimensionalized based on the characteristic time of the borehole field, $t_s = H^2/9\alpha_s$, where H is the borehole length and α_s is the soil thermal diffusivity. D is the depth of the GHE below ground surface and B is the GHE center-to-center spacing.

As currently presented, response factor models are not useful for GHE simulation in the context of WBES without reformulation. The models are formulated in terms of known heat loads on the GHE, which creates a problem for WBES simulation because the GHE loads are not known a priori. Rather, the models are simulated at the same time as the other plant-loop components. The exiting fluid temperature and flow rate from one model is passed downstream to the next model, with all connected components affecting the performance of each other. As a result, the models need to be formulated in terms of known entering fluid conditions and not heat transfer loads. Additional methods also need to be applied to determine the circulating fluid heat transfer rate and associated temperature response instead of the temperature response at the borehole wall.

Response factor methods require that the GHE loads be accounted for from the beginning of the thermal history of the GHE. The result of this is that, as simulation time advances, the amount of effort required to compute the temperature response of the GHE continues to increase. Load aggregation procedures have been developed to mitigate these effects by collapsing loads that occurred far in the past relative to current simulation time be grouped together into one or more blocks. This helps to make response factor method simulation time more amenable to WBES programs. Mitchell & Spitler [26] have reviewed the available load aggregation methods and provided recommendations.

The traditional g-function formulation for short timesteps assumed that the fluid heat transfer rate, q_f, and the borehole wall heat transfer rate, q_b, are approximately equal. The result of this assumption is that the effective borehole wall temperature at short timesteps is negative, which is a non-physical value. Brussieux & Bernier [27] have shown that, as the timestep approaches 0, this value approaches $-2\pi k_s R_b$. In the context of GHE modeling, "short timesteps" are timesteps that are shorter than the transit time of the GHE, which is the time it takes for the circulating fluid to transit through the GHE. Even for common GHE configurations, the transit time can range from minutes to tens of minutes. For a WBES program that can simulate timesteps down to 1 min, accurately computing the short-term transient effects is critical to accurately modeling GHE performance. As a result of these approximations, the outlet fluid temperature computed by the model can be non-physical when high loads are suddenly applied and the simulation timestep is small.

Several authors have modified the original response factor models to improve the short-term, dynamic accuracy of the model. Loveridge & Powrie [28] developed an addition to the original response factor model formulation for modeling concrete pile heat exchangers. The original model was modified by adding a "concrete response function" to account for the transient response of the pile heat exchanger. This method was successfully used to model pile heat exchangers with hourly loads; however, the method was not used for sub-hourly loads. The concrete response function was later used by Alberdi-Pagola [29] and Alberdi-Pagola et al. [30]; however, these studies also appear to be limited to hourly timesteps.

Pile heat exchangers are expected to be relatively short in length when compared to vertical borehole heat exchangers. As a result, the transit time of the circulating fluid from inlet to outlet is also expected to be lower. In addition, because sub-hourly timesteps are not expected, the above model is expected to perform well under these conditions. However, if the simulation timestep approaches the transit time, the model is expected to have trouble predicting the short-term dynamic response. This notwithstanding, the approach shows promise for the current application.

Others have directly computed g-functions for the purpose of directly predicting the GHE exiting fluid temperature. Dusseault & Pasquier [31] briefly mention the method, but a full derivation is given by Pasquier et al. [32]. The approach provides a way to utilize the Eskilson-type g-functions and combine them with the short-timestep g-functions for directly computing the fluid temperature. To compute these entering fluid temperature g-functions, a relatively complicated TRC model is utilized [33,34].

Finally, response factor methods require computing the response factors (i.e., g-functions) themselves. Claesson & Eskilson [23,25] created what is known as the superposition borehole model to generate g-functions. The heat transfer inside the borehole was assumed to be treated separately; therefore, the heat transfer rate computed by the model was from the borehole wall to the surrounding soil. There has been some discussion regarding what boundary conditions were applied at the borehole wall. Cimmino & Bernier [35] investigate this issue in detail by applying an FLS model and concluded that the model applies a uniform borehole wall temperature along the full length of the borehole, and that this temperature is uniform for all boreholes. Libraries of these g-functions representing specific GHE configurations have been published and used for GHE modeling; however, Malayappan & Spitler [36] point out that interpolation between these specific geometries can introduce errors.

ILS and ICS models may also be used to generate g-functions; however, because the internal geometry of the borehole is not considered and because of the various assumptions made regarding the thermal capacity of the borehole, ILS and ICS g-functions are not valid until after a certain time period, which can be up to several hours. Similarly, due to the infinite nature of the models, they will not be accurate at predicting periods when the GHE may interact with the ground surface.

FLS models are commonly used to generate g-functions. Cimmino & Bernier [35] developed a semi-analytical, discretized FLS model to generate response factors which is capable of simulating different borehole wall boundary conditions. Others have developed methods for computing response factors with series or parallel arrangements [16,37,38]. Cimmino [39] later developed a "fast" method for calculating g-functions that takes advantage of symmetry with the borefield to simplify and reduce computations. Cimmino later extended this model to compute g-functions for GHE with series- or parallel-connected boreholes [40,41]. The code for this library is published online [42].

2.3. Thermal Resistance-Capacitance Models

Thermal resistance-capacitance (TRC) models simulate the borehole internal geometry with a simplified resistance-capacitance network, as seen in Figure 2. These models can be formulated to solve for the heat transfer within the borehole by developing a network of temperature nodes that represent the temperature at different locations within the borehole. These temperature nodes are computed by defining an energy balance equation at each node. This energy balance generally also includes a transient term, which accounts for the thermal capacitance of the node. This allows the dynamic evolution of the node temperature over time to be determined.

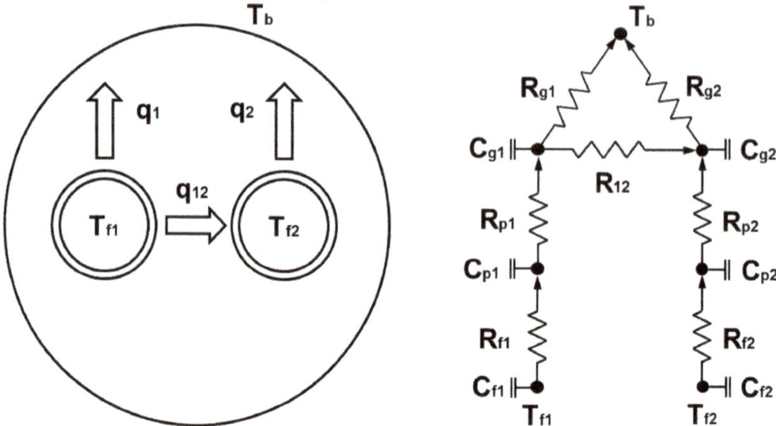

Figure 2. RC-network geometrical representation.

This TRC approach has a benefit in that it does not require direct simulation of all of the geometry within the borehole (as would be done with pure numerical models that use finite-element or finite-volume methods). The result is that the temperature for each node can be solved for by simultaneously solving a set of equations, such as the example in Equation (3), where the left-hand side of the equation represents the transient temperature change with time and the right-hand side represents the energy flows between a node and its neighboring nodes:

$$\rho c_p \frac{dT_i^{t+\Delta t}}{dt} = \sum_{j=1}^{n} \frac{T_j^t - T_i^t}{R_{i,j}} \tag{3}$$

Delta-circuit TRC resistance network models have been developed for single boreholes [43,44]. These models are coupled to several annular regions to complete a radial model of the borehole. The GHE is discretized vertically and multiple segments are stacked and connected which gives a quasi-2D representation of the borehole. Double U-tube and coaxial model configurations have been developed and validated against finite element numerical solutions [45]. Other authors have also added additional resistance-capacitance elements with the TRC model to improve the short-term transient behavior of the models [33,34,46,47].

Ruiz-Calvo et al. [48] describe what is potentially the most promising application of a TRC model for WBES applications. They use the TRC model to compute the heat transfer within the GHE, but then use a response factor model to update the borehole wall boundary temperature. This temperature is updated periodically so that the computations are only performed when needed.

TRC models are useful for representing additional physical detail for GHE modeling without the long simulation time or complexity required to solve a fully discretized numerical simulation. Unfortunately, the simulation time required to compute a TRC model is likely to be longer than would be acceptable for WBES programs. The model by Ruiz-Calvo et al. [46] required 20 s to solve for a 72-h TRT for a single borehole, whereas WBES simulations often run annual simulations of the entire building and connected systems within a similar amount of time. In addition to that, GHEs often need to be simulated for multiple decades to determine whether they have been sized appropriately. Therefore, a direct application of a TRC model within a WBES program would likely increase simulation time significantly, perhaps by several orders of magnitude. This would be considered unacceptable by designers who rely on WBES programs to perform rapid simulations for parametric analysis and design. TRC models are also more challenging to implement due to the complexity of the mathematical methods required. However, TRC models are quite useful for computing g-functions for short timesteps, near or below the transit time.

2.4. Numerical Models

Numerical methods, such as finite element or finite volume methods, have been applied to GHE modeling by several different authors. For example, Al-Khoury [49] developed two pure numerical solutions for modeling GHE. Shao et al. [50] apply these methods to develop a code base. Other pure numerical models have also been developed for simulating GHE performance [51–55]. They are often used as a validation method for other simplified models.

Direct numerical models that fully discretize the spatial domain using finite-volume or finite-element methods are useful for providing detailed results for model validation, but are not useful in the context of GHE modeling for WBES due to the excessive simulation time and inputs required to run such a model. Therefore, they are not considered further in this paper.

3. Methodology

Of the GHE modeling methodologies reviewed and discussed so far, only the response factor models are suitable for WBES applications because of their computational efficiency. This section outlines the development of an enhanced response factor model and discusses how it is used in WBES.

3.1. Enhanced Response Factor Model

Response factor models have been developed and used in WBES by Fisher et al. [56]. That model, which is based on the original formulation by Claesson & Eskilson [23,24], has functioned well; however, as noted previously, it can have trouble simulating short timesteps. So-called "short-timestep" response factors have been developed [27,57,58] to extend these limits, but non-physical behavior can still occur when the timestep is below the transit time.

The original response factor model is given in Equation (4) where the GHE load is used to compute the borehole wall temperature and the GHE mean fluid temperature is computed through a steady-state resistance value, as seen in Equation (5). The g-function values, g, are assumed to be computed at the appropriate time and GHE configuration as seen in Equation (6), where t_n represents the simulation time of the current timestep:

$$T_b = T_s + \sum_{i=1}^{n} \frac{q_i - q_{i-1}}{2\pi k_s} \cdot g \tag{4}$$

$$T_f = T_b + q_n R_b \tag{5}$$

$$g = f\left(\frac{t_n - t_{i-1}}{t_s}, r_b/H, B/H, D/H\right) \tag{6}$$

As presented above, this model is not suited for WBES because it requires the GHE load as an input parameter. However, these equations can be reformulated after assuming a relationship between the mean fluid temperature and the GHE inlet and outlet temperatures, and after assuming that the fluid heat transfer rate, q_f, and the borehole wall heat transfer rate, q_b, are equal. See Figure 3 for a simplified schematic of a borehole, with locations indicated for q_f and q_b. A full derivation of this reformulated model has been provided by Mitchell [59].

The approach adopted in this work may be considered a blend between the approach by Loveridge & Powrie [28] and Pasquier et al. [32]. Specifically, Loveridge & Powrie developed a response factor model with a separate g-function for modeling the short-term response of a concrete pile heat exchanger, and the model given by Pasquier et al. developed a way to combine the Eskilson-type g-functions with exiting fluid temperature g-functions for directly computing the exiting fluid temperature.

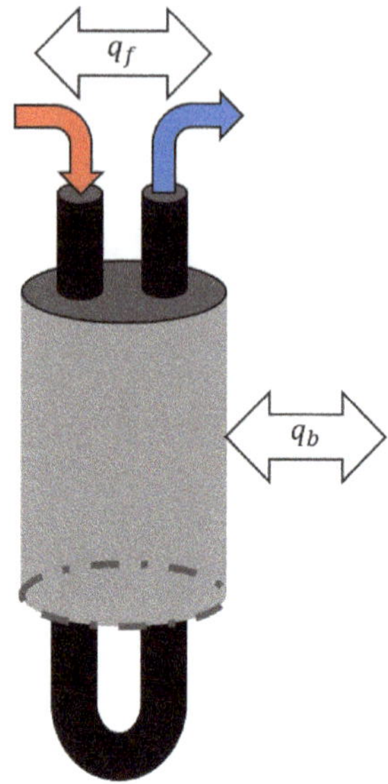

Figure 3. Single U-tube borehole.

The proposed model is given in Equation (7). In this formulation, the GHE exiting fluid temperature is computed directly by modifying the historical response factor model (Equation (5)) with the addition of so called "exiting fluid temperature" (ExFT) response factors. These ExFT response factors are computed to determine the temperature difference of the GHE exiting fluid temperature from the borehole wall temperature. As a result, the model can be thought of as three individual parts, given in Equations (8)–(10). Part 1 (Equation (8)) computes the temperature difference between the borehole wall and the far-field soil; part 2 (Equation (9)) computes the temperature difference between the outlet fluid temperature and the borehole wall temperature; and part 3 (Equation (10)) computes the fluid heat transfer rate. All three parts are solved together simultaneously:

$$T_{out} = T_s + \sum_{i=1}^{n} \frac{q_i - q_{i-1}}{2\pi k_s} \cdot g + R_b \sum_{i=1}^{n} (q_i - q_{i-1}) \cdot g_b \tag{7}$$

$$T_b - T_s = \sum_{i=1}^{n} \frac{q_i - q_{i-1}}{2\pi k_s} \cdot g \tag{8}$$

$$T_{out} - T_b = R_b \sum_{i=1}^{n} (q_i - q_{i-1}) \cdot g_b \tag{9}$$

The first part comes from the historical response factor model. By applying this equation, the GHE borehole wall temperature difference from the soil temperature can be computed. The second part is the new formulation that computes the GHE exiting fluid temperature difference from the GHE borehole wall temperature.

To clarify, the heat transfer rate applied here is the calorimetric fluid heat transfer rate as calculated from the inlet and outlet temperatures as well as the flow rate of the GHE. This is seen in Equation (10), where T_{in} is the GHE entering fluid temperature and T_{out} is the GHE exiting fluid temperature:

$$q_f = \frac{\dot{m}_f c_{p,f}}{H_{tot}} (T_{in} - T_{out}) \tag{10}$$

There are several reasons for formulating the model as shown:

- Significant effort has been expended to understand and improve the original response factor model. Since its publication, researchers have performed many studies that enhance understanding of the methods and improve on the original work. This work similarly builds on the original model, about which much is already known, and which has already been widely adopted.
- Because the formulation builds on the historical response factor model, software or other programs that already have this model implemented can more easily make modifications to incorporate the enhancements. This simplifies adoption of the model in WBES environments.
- Again because the formulation builds on the historical response factor model, methods for generating standard borehole wall temperature g-functions are still applicable (as are load aggregation procedures, which are critical to maintaining low simulation times).
- By clearly defining the heat transfer rate applied in this model as the fluid's heat transfer rate, the domain inside the borehole and the domain between the borehole wall and the far-field soil temperature can be coupled together easily through two separate response factor computations. This more easily allows the transient effects to be handled, even down to timesteps below the transit time of the GHE circulation fluid.

The following section provides additional details about how the original and ExFT g-functions were developed and how the model is used in WBES.

3.2. Model Reformulation for WBES Usage

In order for the model to be applied in WBES, the model needs to be formulated from known quantities. In this case, we wish to pass an entering fluid temperature and mass flow rate to the GHE model and have it compute and return the exiting fluid temperature.

Starting with Equation (7), we will let Equation (11) apply, so the current model becomes that shown in Equation (12):

$$c_0 = \frac{1}{2\pi k_s} \tag{11}$$

$$T_{out} = T_s + c_0 \sum_{i=1}^{n} (q_i - q_{i-1}) \cdot g + R_b \sum_{i=1}^{n} (q_i - q_{i-1}) \cdot g_b \tag{12}$$

To clarify, in response factor models, the multiplication of the heat rate differences by the g-function values is a convolution operation. For example, if three timesteps have occurred, the summation-convolution operation would be as follows:

$$
\begin{aligned}
\sum_{i=1}^{3} (q_i - q_{i-1}) \cdot g = {} & (q_3 - q_2) \, g \left(\frac{t_3 - t_2}{t_s} \right) \\
& + (q_2 - q_1) \, g \left(\frac{t_3 - t_1}{t_s} \right) \\
& + (q_1 - q_0) \, g \left(\frac{t_3 - t_0}{t_s} \right)
\end{aligned}
\tag{13}
$$

Next, because the current heat load is not known, it is deconvolved from the response factor summations:

$$
\begin{aligned}
T_{out} = {} & T_s + c_0 \, (q_n - q_{n-1}) \cdot g_n + c_0 \sum_{i=1}^{n-1} (q_i - q_{i-1}) \cdot g \\
& + R_b \, (q_n - q_{n-1}) \cdot g_{b,n} + R_b \sum_{i=1}^{n-1} (q_i - q_{i-1}) \cdot g_b
\end{aligned}
\tag{14}
$$

For simplification, we will let the g-function values for the current heat load be noted as shown in Equations (15) and (16):

$$g_n = g\left(\frac{t_n - t_{n-1}}{t_s}\right) \tag{15}$$

$$g_{b,n} = g_b\left(\frac{t_n - t_{n-1}}{t_s}\right) \tag{16}$$

From there, the historical terms are gathered together in Equation (17), which then leaves Equation (18):

$$c_1 = c_0 \sum_{i=1}^{n-1} (q_i - q_{i-1}) \cdot g + R_b \sum_{i=1}^{n-1} (q_i - q_{i-1}) \cdot g_b \tag{17}$$

$$T_{out} = T_s + c_0 (q_n - q_{n-1}) \cdot g_n + R_b (q_n - q_{n-1}) \cdot g_{b,n} + c_1 \tag{18}$$

After this, we apply another replacement to simplify the notation (Equation (19)), which then leaves Equation (20) after the current heat transfer rate, q_n, and the previous timestep heat transfer rate, q_{n-1}, are gathered together:

$$c_2 = c_0 g_n + R_b g_{b,n} \tag{19}$$

$$T_{out} = T_s + c_2 q_n - c_2 q_{n-1} + c_1 \tag{20}$$

Because we wish to clearly define the model in terms of the fluid heat transfer rate, Equation (10) is again introduced in Equation (21), after another simplification is applied in Equation (22):

$$q_f = c_3 (T_{in} - T_{out}) \tag{21}$$

$$c_3 = \frac{\dot{m}_f c_{p,f}}{H_{tot}} \tag{22}$$

After Equation (21) is substituted into Equation (20) and after some algebraic manipulation, we arrive at the final solution in Equation (23):

$$T_{out} = \frac{T_s + c_2 c_3 T_{in} - c_2 q_{n-1} + c_1}{1 + c_2 c_3} \tag{23}$$

The model is now cast in terms of known quantities. The question remains, though, regarding how to compute the g-function values. To do this, a simplified dynamic borehole model has been developed. This will be discussed in the next section, along with computation of the g-function values.

3.3. Dynamic Borehole Model

As discussed previously, this model can be considered a blend between the approaches outlined by Loveridge & Powrie [28] and Pasquier et al. [32]. In Loveridge & Powrie's paper on pile heat exchangers, the authors used a detailed numerical model to compute the transient response [60]; in Pasquier et al.'s work on the borehole heat exchanger, a detailed TRC model was used to determine the short-term, dynamic response [33,34]. A similar approach will be taken here; however, because computation time is so important, a fast solution is desired.

To do this, a simple dynamic borehole model has been developed, which will be used to compute the ExFT g-functions. The model is composed of two simple, 1D dynamic pipe models for simulating the transit delays and a simple five-node TRC model for simulating heat transfer within the borehole. This is seen in Figure 4.

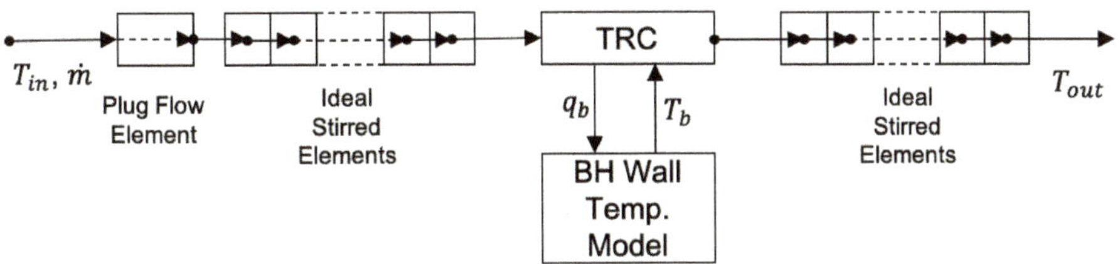

Figure 4. Dynamic model borehole schematic.

3.3.1. Simple Borehole TRC Model

A schematic of a segment of the simple TRC model borehole is shown in Figure 5. The model is a simple, five-node model. For comparison, the model used by Pasquier & Marcotte [34] has 16 nodes per segment with multiple segments stacked together. The number of segments used was left as a parameter that could be varied during model validation.

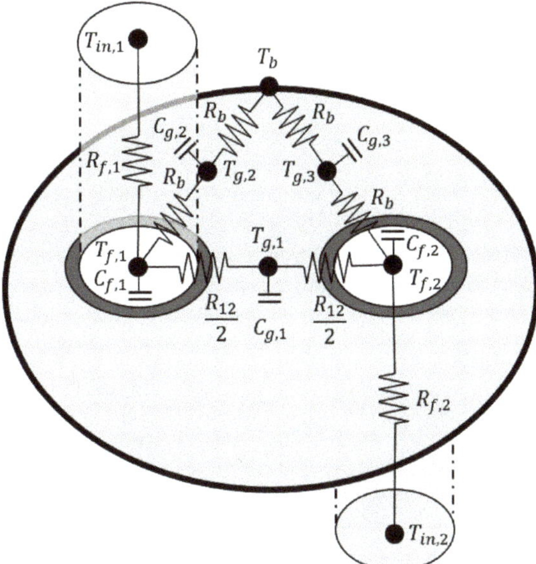

Figure 5. Simple TRC model.

The resistance values are set using the first-order approximation of multipole method [61–63], where R_b is the borehole thermal resistance and R_{12} is the direct-coupling resistance. Javed & Spitler [64] discuss the accuracy of these methods and note that, in most situations, the first-order approximation is accurate to within 1%.

Equations for computing R_b have been given in Javed & Spitler [64,65]. Direct-coupling resistances using the multipole method are not given in the literature, so it has been derived from the Delta-circuit resistance network and given in Equation (24) in terms of the borehole resistance and the total internal borehole resistance, R_a (both of which can be computed directly from the multipole method). Fluid resistance in the flow direction is computed as shown in Equation (25):

$$R_{12} = \frac{4R_a \cdot R_b}{4R_b - R_a} \tag{24}$$

$$R_f = \frac{1}{\dot{m}_f c_{p,f}} \tag{25}$$

Equations for computing the temperature of each of the temperature nodes from the borehole segment model are given below.

$$C_{f,1}\frac{dT_{f,1}}{dt} = \frac{T_{in,1} - T_{f,1}}{R_{f,1}} + \frac{T_{g,1} - T_{f,1}}{R_{12}/2}dz + \frac{T_{g,2} - T_{f,1}}{R_b}dz \tag{26}$$

$$C_{f,2}\frac{dT_{f,2}}{dt} = \frac{T_{in,2} - T_{f,2}}{R_{f,2}} + \frac{T_{g,1} - T_{f,2}}{R_{12}/2}dz + \frac{T_{g,3} - T_{f,2}}{R_b}dz \tag{27}$$

$$C_{g,1}\frac{dT_{g,1}}{dt} = \frac{T_{f,1} - T_{g,1}}{R_{12}/2}dz + \frac{T_{f,2} - T_{g,1}}{R_{12}/2}dz \tag{28}$$

$$C_{g,2}\frac{dT_{g,2}}{dt} = \frac{T_{f,1} - T_{g,2}}{R_b}dz + \frac{T_b - T_{g,2}}{R_b}dz \tag{29}$$

$$C_{g,3}\frac{dT_{g,3}}{dt} = \frac{T_{f,2} - T_{g,3}}{R_b}dz + \frac{T_b - T_{g,3}}{R_b}dz \tag{30}$$

The thermal capacity for each node is set from the product of their respective heat capacity and volume. This is shown below for the fluid capacity elements, assuming equivalent pipe sizes for both pipes in the segment:

$$C_{f,1} = \rho_f c_{p,f} V_{f,1} \tag{31}$$
$$C_{f,2} = C_{f,1} \tag{32}$$

The grout capacity nodes are slightly different because it is difficult to determine what fraction of the total volume grout should be associated with each node. Therefore, this fractional value is included in the grout capacity nodes, which could be adjusted during model validation. Total grout volume can be determined by taking the total volume inside of the borehole and subtracting the total pipe volume:

$$C_{g,1} = f_g \rho_g c_{p,g} V_g \tag{33}$$

$$C_{g,2} = C_{g,3} = \frac{(1 - f_g)\rho_g c_{p,g} V_g}{2} \tag{34}$$

Equations (26) through (30) are solved simultaneously using a Runge–Kutta fourth-order time integration scheme [66]. Model boundary conditions are the inlet temperatures and mass flow rates for each leg of the U-tube segment as well as the borehole wall temperature. All node temperatures are initialized to the undisturbed ground temperature.

The borehole wall boundary temperature is updated each timestep using the response factor calculation seen in Equation (4). g-function response factors are computed using the pygfunction library [42], which uses FLS methods for long-timestep g-functions [39]. Short-timestep g-functions are computed using the 1D, radial finite volume model developed by Xu & Spitler [58]. However, instead of using the model to evaluate the fluid temperature rise and then subtract the borehole resistance—which effectively results in a negative borehole wall temperature response at short timesteps—the actual borehole wall temperature rise is evaluated and then the short-timestep g-functions are computed.

3.3.2. Dynamic Pipe Model

In order to simulate the transit delay effects within the borehole, a dynamic pipe model is used. This dynamic pipe model was originally developed by Bishoff & Levenspeil [67] and later improved by Skoglund & Dejmek [68]. Rees [69] applied the model in a fashion similar to the current application, except the author used a 2D, radial-angular finite volume model to compute the heat transfer within the borehole. The model applies a single plug-flow element that tracks the history of the pipe inlet temperatures, then the temperature nodes along the pipe are mixed together in ideally stirred tank elements. Skoglund & Dejmek [68] recommend 16 elements to achieve a good balance between accuracy and performance. In the model applied here, only the plug-flow time delay is applied to the inlet pipe. The outlet pipe directly mixes the element temperatures together to determine the dynamic temperature response. Computing the temperature of the stirred elements can be accomplished with a tri-diagonal matrix inversion solution, often referred to as the Thomas Algorithm [66]. This convenient formulation allows the computations to execute quickly.

The model is limited to simulating turbulent flows; however, the present study also used the model for laminar flows, so addressing this is recommended for future work. Note that the general methods developed in this section for an enhanced response factor model are not dependent on this particular dynamic borehole model. Any dynamic model that computes the short-term, transient temperature response of the borehole could be used, assuming it is accurate and fast to compute.

3.3.3. Dynamic Borehole Model Validation

In order to validate the dynamic borehole model, a parametric study using the model was run and compared against experimental data collected by ourselves in 2014. The data were collected from two multi-flowrate thermal response tests (MFRTRT) where the heat input rate was held constant and the flow rate was varied at discrete intervals. The first "high" flow rate test varied the flow from 0.15 kg/s, to 0.30 kg/s, to 0.45 kg/s. This resulted in flows with Reynolds numbers of approximately 12k, 21k, and 32k, respectively. The second "low" flow rate test varied the flow from 0.02 kg/s to 0.2 kg/s in eight intervals. This resulted in flows with Reynolds numbers ranging from approximately 2.2k to 22k. Additional information regarding these tests is given by Beier et al. [70]. Parameters regarding the borehole configuration match those provided by Beier et al.

Because the dynamic borehole model is only expected to run long enough to generate the ExFT g-functions, the model was only run for 24 h using the GHE inlet conditions logged from the experimental data. The parameters varied in the study were the grout fraction, number of iterations, number of TRC model segments, and the simulation timestep. The results from the parametric study plot the simulation time required for exercising the model for a single flow rate on the vertical axis, and plot the root mean squared error of the exiting fluid temperature on the horizontal axis.

Figure 6 shows the data with the Pareto front identified. The Pareto front was identified using the "pareto" Python library [71], which employs the popular non-dominated sorting genetic algorithm NSGA-II [72].

The Pareto data are given in Table 1. The second configuration identified was then used to run the dynamic borehole model to simulate using additional data from the MFRTRT. The mean bias error is also given, which is the average exiting fluid temperature error. Even at 7.8 s, the time required to compute ExFT g-functions is acceptable for WBES. Note that the software used for this project was implemented in the Python programming language on a Windows 10 computer with a 3.4 GHz processor. Python is not known for its computational speed, so these times could be improved significantly by implementing the code in a compiled programming language, such as C++.

Table 1. Dynamic borehole model Pareto data (note: MBE = mean bias error).

N_{it}	N_{seg}	Δt	f_g	RMSE [C]	MBE [C]	Time [s]
3	1	60	0.75	0.229	−0.073	8.5
2	1	60	0.5	0.322	−0.169	7.8
1	1	60	0.5	0.634	−0.448	6.9

Figure 7 shows the full nine days of the high-flow MFRTRT. The model is driven from the experimental measurements of the GHE inlet temperature and flow rate. The outlet temperature error is also plotted. Despite the dynamic borehole model being extremely simple, the results are promising. Some error associated with transit time computation is shown near the beginning of the test; however, the general trend of the model follows the experimental data well. For most of the first two hours, the exiting fluid temperature error is within $0.5\,^\circ$C.

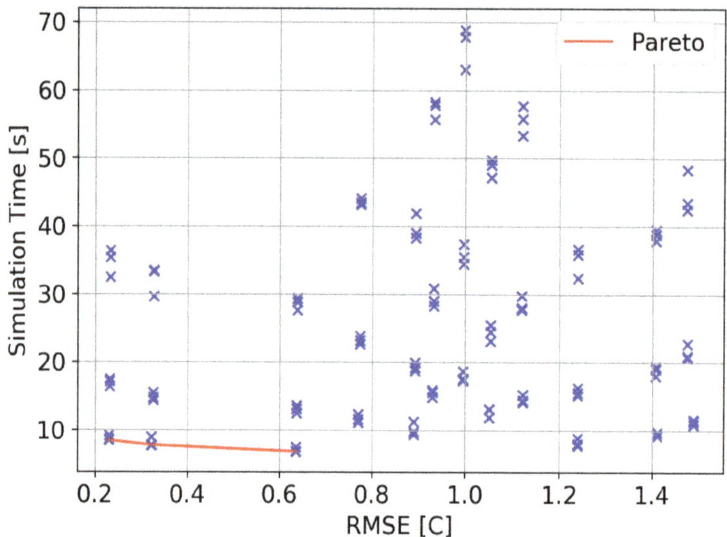

Figure 6. Pareto values for variations in dynamic borehole model input parameters (note: RMSE = root mean squared error).

Again, during most of the test, the exiting fluid temperature error is within $0.5\,^\circ$C, with the exception of the periods during the flow rate step-change. Again, these are caused by errors in the transit delay prediction.

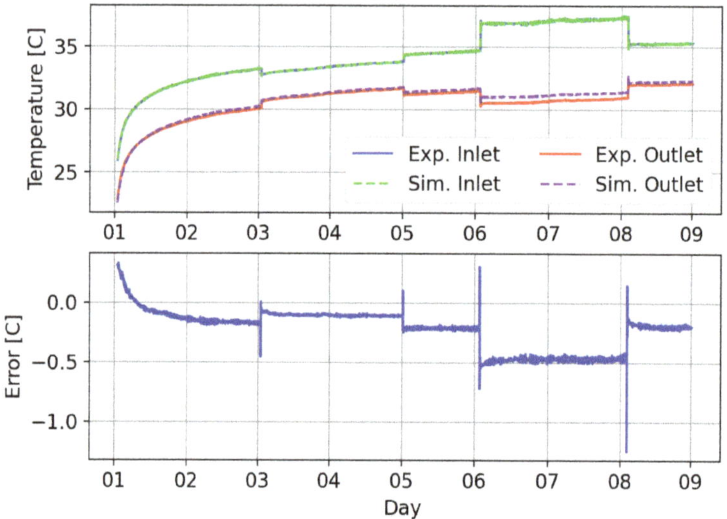

Figure 7. Dynamic borehole model compared to high-flow MFRTRT.

Figure 8 shows the first two days of the low-flow MFRTRT. Some experimental interruptions occurred, which require some special treatment in order to apply the current model. Therefore, no additional comparisons are made beyond the two days shown. Exiting fluid temperature error is slightly higher than the high-flow MFRTRT.

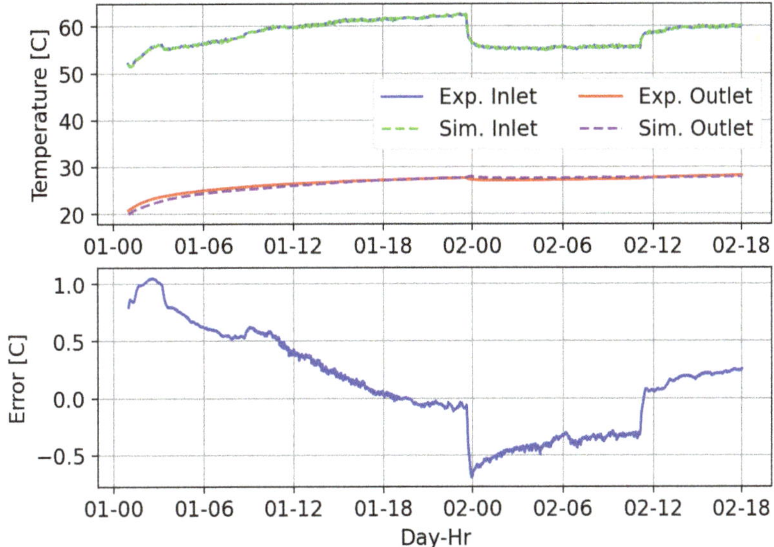

Figure 8. Dynamic borehole model compared to 2 days of low-flow MFRTRT.

Improvements could certainly be made to the dynamic borehole model; however, the model is sufficiently accurate for generating ExFT g-functions for the enhanced response factor model.

3.4. Exiting Fluid Temperature Response Factor Generation

To generate the ExFT g-functions, the simplified dynamic borehole model is exercised with a constant heat load and constant flow rate. Once the model is initialized, the flow rate is held constant and the inlet temperature is computed based a first-law energy balance for a fixed heat input rate. The dynamic borehole model exiting fluid temperature is then fed back into the inlet of the GHE after again applying a first-law energy balance for a fixed heat input rate. By exercising the model in this fashion, the transit delay effects can be estimated and incorporated into the g_b values computed.

The temperature response data from the model is then converted to the ExFT g-function values and given for their respective non-dimensional time. The ExFT g-functions are computed as shown in Equation (35). Non-dimensional time that is used for response factor models is computed as shown in Equation (36):

$$g_b = \frac{T_{ExFT} - T_b}{q_f \cdot R_b} \tag{35}$$

$$\ln\left(t/t_s\right) = \ln\left(\frac{t}{H_{ave}^2 / \left(9\alpha_s\right)}\right) \tag{36}$$

In Equation (35), the temperature difference and the heat transfer rate are expected to be proportional to one another, so the only changes expected in the resulting g_b values computed will be due to the borehole resistance calculation. R_b includes the pipe resistance, which will be affected by changes in flow rate and fluid temperature. Because R_b can change significantly with the flow rate, it is expected that the flow rate values used for computing g_b are not arbitrary, and that g_b will need to be computed for multiple values in order to account for variations in flow rate.

Figure 9 shows the g_b values which were computed plotted vs. non-dimensional time. As expected, higher flow rate data begin to increase sooner than low flow rate data due to the differences in the transit time from different flow rates. Undulations in the data are directly related to the transit delay effects. Excerpted data from the g_b computations at $t = 24\,\text{h}$ are given in Table 2.

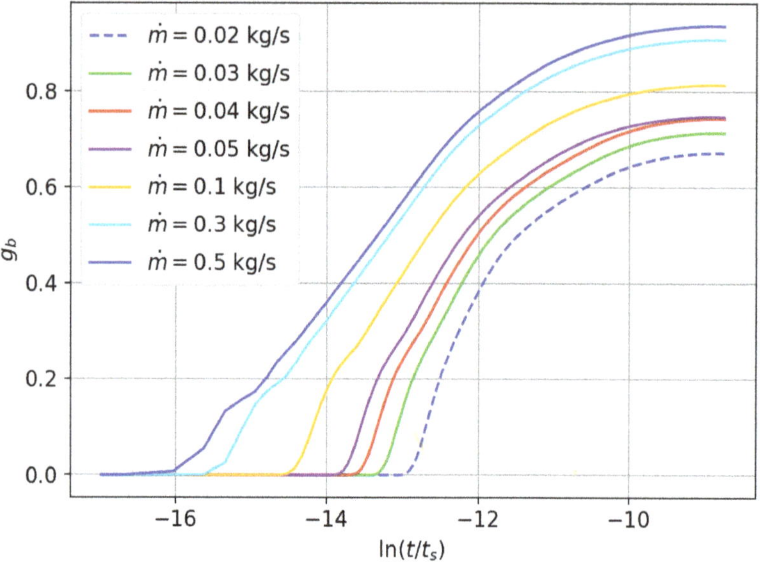

Figure 9. g_b vs. $\ln\left(t/t_s\right)$.

From the table, we can see that the fluid heat transfer rate is nearly constant. Additionally, the borehole resistance decreases with flow rate, except for the first flow rate of 0.02 kg/s which is 0.5% lower than the resistance at 0.03 kg/s. The cause of this is unknown at the moment but is likely attributable to differences in fluid property computations due to temperature differences. For reference, fluid properties were computed using the CoolProp thermophysical properties library [73]. The borehole wall temperature is also nearly constant as is expected because the heat transfer expected at the borehole wall should approach a steady value after enough time, regardless of the fluid flow rate.

Table 2. Data from g_b calculations taken at $t = 24\,\text{h}$.

$\dot{m}_f$	Re	$\ln\left(t/t_s\right)$	g_b	T_{out}	T_b	q_f	R_b	$T_{out} - T_b$	$q_f \cdot R_b$
kg/s	-	-	-	°C	°C	(W/m)	°C/(W/m)	°C	°C
0.02	1400	−8.8	0.67	18.80	17.34	9.989	0.216	1.45	2.16
0.03	1967	−8.8	0.71	18.89	17.35	9.991	0.217	1.55	2.16
0.04	2534	−8.8	0.74	18.92	17.35	9.993	0.211	1.57	2.11
0.05	3079	−8.8	0.75	18.62	17.35	9.994	0.169	1.27	1.69
0.1	5894	−8.8	0.81	18.62	17.36	9.996	0.154	1.26	1.54
0.3	17,209	−8.8	0.91	18.73	17.36	9.996	0.151	1.37	1.51
0.5	28,535	−8.8	0.94	18.76	17.36	9.994	0.150	1.40	1.50

3.5. Borehole Wall Temperature Response Factor Generation

Standard g-function values used for computing the borehole wall temperature rise could be computed in a number of ways. These have been discussed in the literature review and are not repeated here. The methods developed by Cimmino [39] will be used through the pygfunction Python library [42] for computing the long-timestep g-functions. Short-timestep borehole wall temperature g-functions are computed using the 1D radial finite-volume model developed by Xu & Spitler [58]. These g-functions are plotted in Figure 10 for the MFRTRT borehole discussed in Beier et al. [70]. Because it can be difficult to understand non-dimensional time values, some dimensional values are given as well.

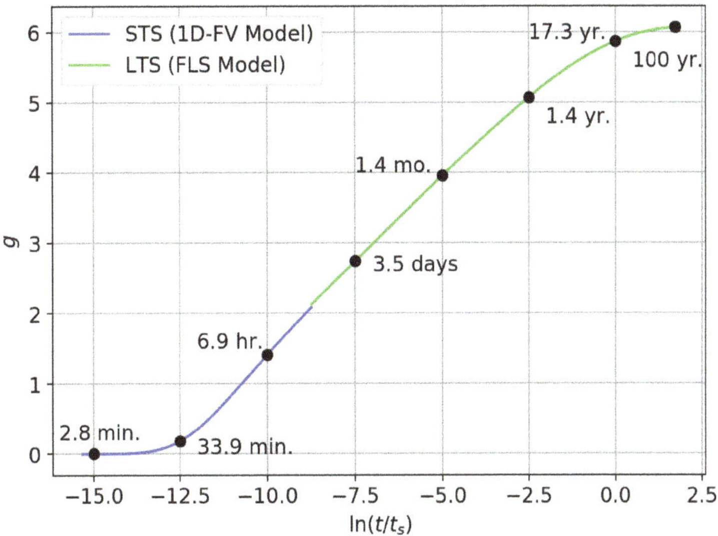

Figure 10. g vs. $\ln\left(t/t_s\right)$.

As stated previously, the g-function values are evaluated to give the borehole wall temperature rise, not the fluid temperature rise. For this reason, the values approach 0 for lower $\ln\left(t/t_s\right)$ values. The original implementation of the model by Xu & Spitler [58] subtracts the steady-state borehole resistance so the mean fluid temperature could be computed. Other short-timestep g-function generation methods do this as well; however, this is not correct for the current model's formulation. As expected, some time is required for the heat to conduct from the fluid, through the grout, to the borehole wall. As seen in the plot, this time is on the order of 20–30 min for this particular borehole.

The model by Xu & Spitler [58] is a 1D, radial finite-volume model as outlined by Patankar [74], which is formulated to be solved using a tri-diagonal solution scheme. The solution is computed with a fixed timestep of 2 min for 24 h, after which the long-timestep g-function methods are applied.

As with the dynamic borehole model, the choice of flow rate will affect the short-timestep borehole wall g-functions. However, because we are evaluating the borehole wall temperature with a fixed heat input rate, the effect on the short-timestep g-functions is expected to be quite small, and therefore the choice of flow rate used to compute the short-timestep and long-timestep g-functions is arbitrary.

3.6. Methodology Summary and Discussion

The methods used in this enhanced model are similar to previous response factor models; however, the method has been modified to better compute the GHE exiting fluid temperature at short timesteps. To simplify, the methods described previously could be summarized and implemented as follows:

1. *Long-timestep borehole wall temperature g-functions* could be computed with any number of
    approaches outlined previously. In this work, an FLS approach was used by utilizing the
    pygfunction library by Cimmino [42]. No modifications are required to the methodology used
    to compute these g-functions. Cimmino has produced significant quantification regarding the
    speed of these methods [39] and has shown that for borefields with up to 64 boreholes—which is
    realistically expected to capture most of the WBES use cases—g-functions can be computed
    in times from a few seconds to a few minutes. Note that the code by Cimmino is written in
    Python; the computation time could likely be reduced to some extent after being implemented in
    a compiled programming language.

2. *Short-timestep borehole wall temperature g-functions* could also be computed using any accurate
    borehole model that captures the dynamic effects of the borehole thermal capacity on the
    borehole wall temperature. In this work, the model by Xu & Spitler [58] was used because
    it uses a simplified geometry that accounts for the borehole thermal capacity. In addition,
    it is a 1D, radial finite-volume model that can be solved rapidly using a tri-diagonal matrix
    formulation. Note that the original model has to be slightly modified to compute the temperature
    at the borehole wall, and not at the fluid. This model is highly efficient and can compute the
    short-timestep g-functions in a matter of seconds.

3. *Exiting fluid temperatures g-functions* are computed with a simplified dynamic borehole model
    that accounts for the transit delay of the circulation fluid via utilization of a transit delay pipe
    model. As before, the method presented in this paper does not rely on this specific dynamic
    GHE model; rather, any dynamic GHE model that accurately captures the transit delay effects
    of the borehole could be used, assuming that it is sufficiently fast and accurate. The borehole
    wall boundary temperature was updated by computing the heat flux at the borehole wall and
    then using that information to inform the original heat-input-formulated response factor model
    (Equation (4)). The borehole wall temperature resulting from that was then set as the borehole
    wall temperature boundary condition and updated at each timestep. Note that the short-timestep
    g-functions computed in the previous step are used here for determining the updated borehole
    wall boundary temperature. The dynamic model used here was able to compute the exiting fluid
    temperature g-functions for a single flow rate in 7.8 s, which is an acceptable time for WBES.
    Note that this should be repeated for different flow rates and that additional work should be
    done to determine how many different g_b values are needed.

Though the method does require some computation time for the various required g-functions,
these times are acceptable for WBES applications. Additionally, if the GHE configuration does
not change, these values could be cached and stored for reuse during subsequent simulations.
Provided that the GHE is not changed, these values do not need to be recomputed for each simulation.

4. Validation

The enhanced response factor model is validated using the MFRTRT data. Validation results for
the high- and low-flow tests are presented in this section. For all cases, the experimental measurements
for temperature and flow rate were used as the inlet conditions for the GHE model.

4.1. High-Flow MFRTRT

Figure 11 shows the temperature comparison for the first two hours of the high-flow MFRTRT.
The simulation was run using a 60-s timestep. As occurred during the comparison of the dynamic
borehole model, some transit delay errors are present during the first five minutes of the simulation,
which result in some error in the temperature predicted. However, once the first few transit periods
have passed, the model and experimental data agree quite well.

Figure 12 shows the temperature comparison for the first flow rate change using a 60-s timestep.
During this period, the flow rate is changed from approximately 0.3 kg/s up to 0.45 kg/s, which results
in a smaller temperature difference between the inlet and outlet of the GHE. The figure shows good

a dynamic response compared to the experiment during this step change in flow, and the absolute outlet temperature error remains under 0.2 °C.

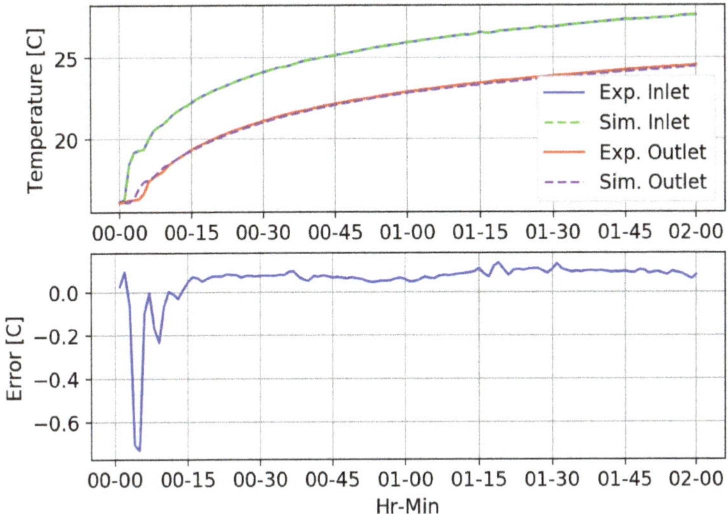

Figure 11. Enhanced model results for high-flow MFRTRT. First 2 h. 60-s timestep.

Figures 13 and 14 show the full high-flow MFRTRT data compared using the enhanced response factor model using 60-s and 1-h timesteps, respectively. Figure 13, as in previous plots, shows higher errors at times when the flow rate changes, but level off after that. Again, these errors are attributed to the errors in the transit time predicted using the dynamic borehole model. The overall temperature difference is generally under predicted by less than 0.5 °C.

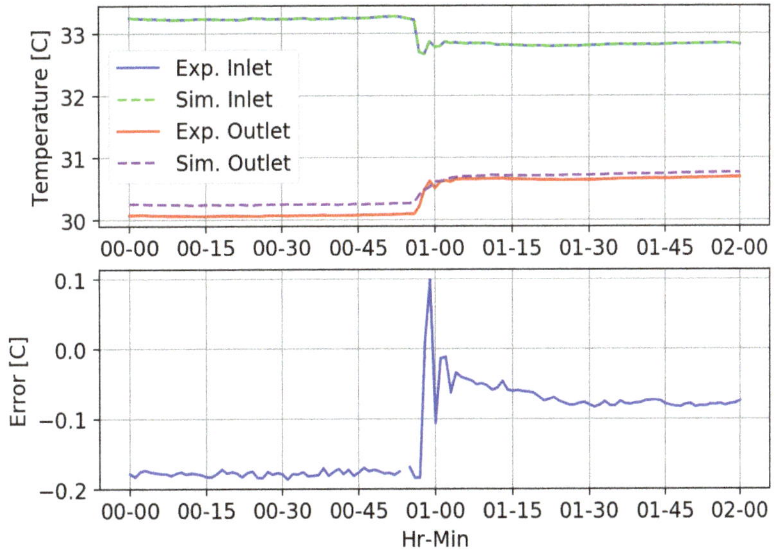

Figure 12. Enhanced model results for high-flow MFRTRT. First flow rate change. 60-s timestep.

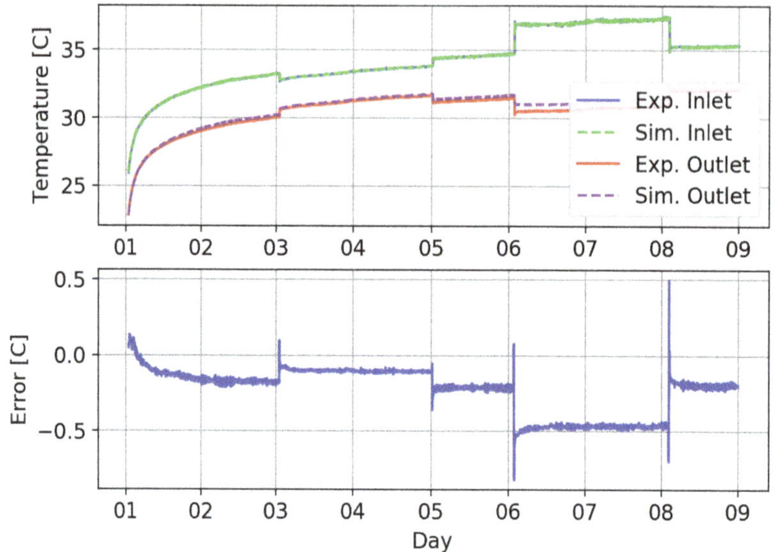

Figure 13. Enhanced model results for high-flow MFRTRT. Full data set. 60-s timestep.

For the simulation using a 1-h timestep shown in Figure 14, these transit time errors are not present because a 1-h timestep is 10–30× the GHE transit time. The overall trend of the exiting fluid temperature error is consistent with the 60-s timestep simulation data.

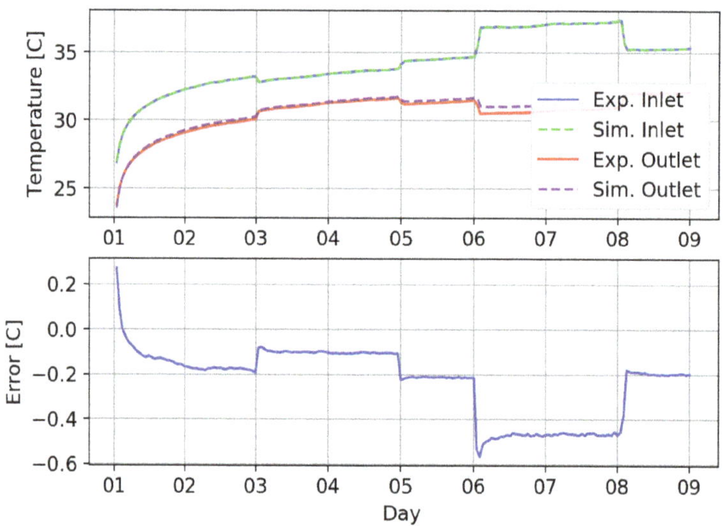

Figure 14. Enhanced model results for high-flow MFRTRT. Full data set. 1-hr timestep.

Figures 15 and 16 show the heat transfer error from the experimental data and the simulation, for a 60-s and 1-h timestep, respectively. In both cases, the heat transfer error is within 10%, except during flow changes, and near 5–6% for the test duration. Note that the experimental uncertainty in the heat transfer measurements is not shown here but is discussed in Beier et al. [70]. For the high-flow test, the heat transfer error is generally around ±3–7%, depending on flow rate. Additionally, note that the soil and grout parameters are estimated values, and that changes in these values can easily affect the results enough to affect the error predicted. Further optimization of the soil and grout parameter values could result in reductions to the overall error.

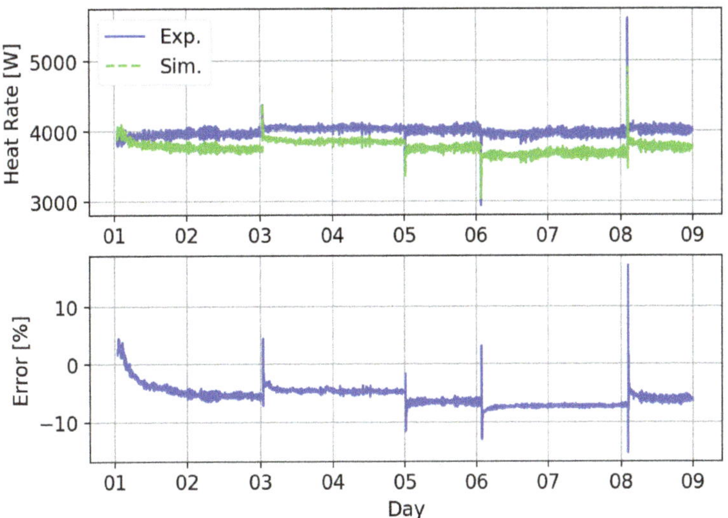

Figure 15. Enhanced model heat transfer results for high-flow MFRTRT. Full data set. 60-s timestep.

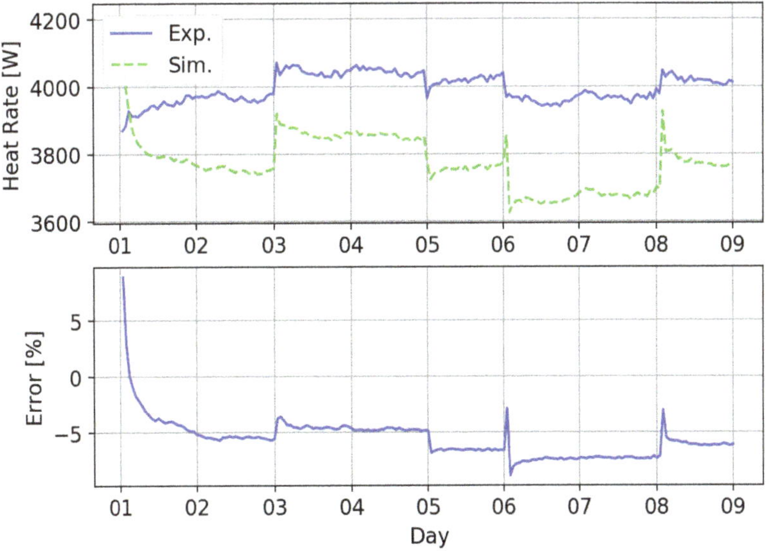

Figure 16. Enhanced model heat transfer results for high-flow MFRTRT. Full data set. 1-hr timestep.

4.2. Low-Flow MFRTRT

Figure 17 shows the model compared to 68 h of data from the low-flow MFRTRT. The absolute exiting fluid temperature error is higher than the high-flow test, as the temperature difference is under predicted by about 1 °C. As with the high-flow test, a similar error can be seen from in Figure 8 from the direct simulation using the dynamic borehole model. Again, this suggests that errors in the enhanced response factor method can be addressed by improving the accuracy of the dynamic borehole model used to generate the ExFT g-functions.

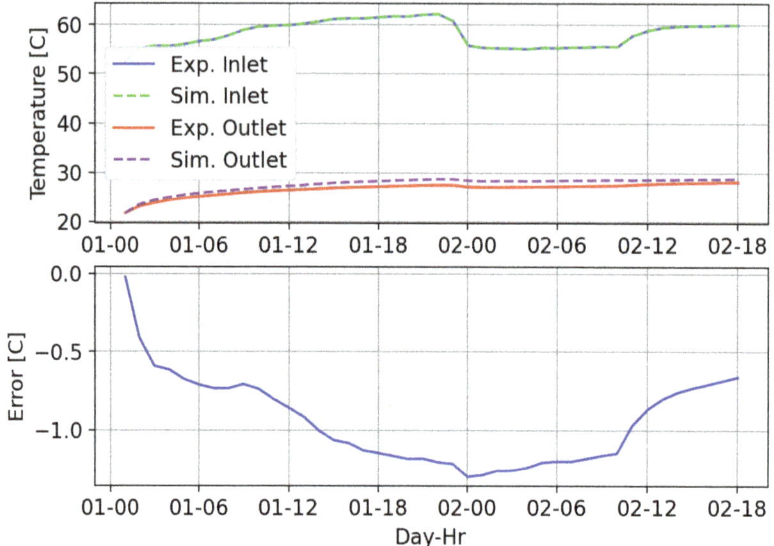

Figure 17. Enhanced model temperature results for low-flow MFRTRT. Full data set. 1-hr timestep.

Figure 18 shows the heat transfer results for the low-flow test compared to the simulation. As before, the experimental heat transfer rate errors are not plotted, but, for this particular flow rate, they are around ±3%. Variations in soil properties will also affect the results, and in practice the inlet temperatures and heat transfer rates will be adjusted naturally during an overall system simulation.

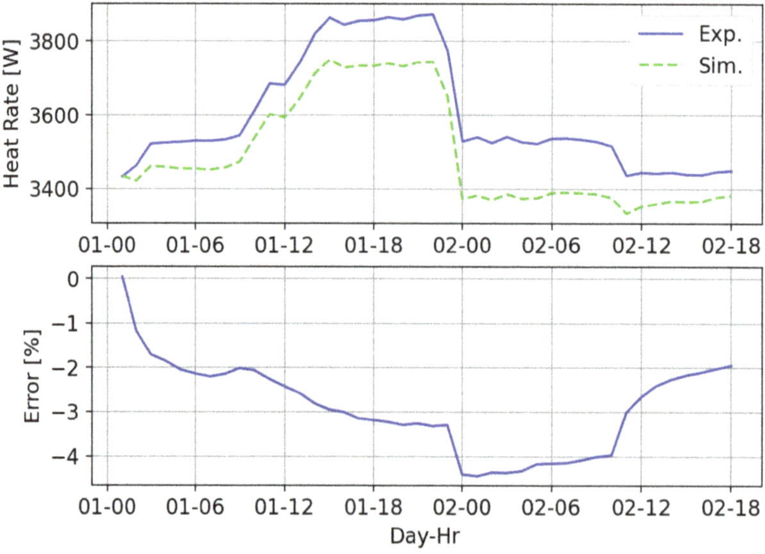

Figure 18. Enhanced model heat transfer results for low-flow MFRTRT. Full data set. 1-hr timestep.

5. Conclusions

This work describes an enhanced response factor model for use in whole-building energy simulation environments. This model combines long-timestep and short-timestep g-functions that give responses at the borehole wall, with the exiting fluid temperature g-functions proposed by Pasquier et al. [32]. Using the new model, our research has shown that accurate dynamic behavior can be simulated for GHE using response factor calculations. The results are shown to be accurate for timesteps that are shorter than the GHE transit time, as well as for long timesteps.

A simplified dynamic borehole model was also developed, which is used to generate exiting fluid temperature g-functions. A Pareto front analysis was performed to determine the optimized model parameters for speed and accuracy. Computation time for the exiting fluid temperature g-functions was shown to be computed in less than 8 s on a desktop PC while keeping the root mean squared error below 3 °C. Inaccuracies in the enhanced response factor model were shown to be directly related to inaccuracies in the simplified dynamic borehole model, though the methods generally compare well against experimental MFRTRT measurements.

The new method requires computation of long-timestep, short-timestep, and exiting fluid temperature g-function values; however, the methods utilized here are efficient and are acceptable for WBES applications. Given some intelligent programming, the g-functions would only need to be computed one time for each unique GHE configuration. If the GHE configuration is not changed between simulations, the g-function values could be saved and reused during subsequent simulations.

Although this work has shown that the model is sufficiently fast and acceptably accurate, further research in these areas might improve accuracy, computational speed, and applicability to different ground heat exchanger types:

1. The pipe model currently used does not account for laminar flow; therefore, the pipe model should be enhanced to accurately account for these conditions.
2. Any dynamic borehole model could be used to generate ExFT g-functions; therefore, a study should be performed to assess this model along with other existing models for accuracy and performance.
3. The methods described here could also be used to model double U-tube or coaxial ground heat exchangers. Dynamic borehole models for these configurations would be needed in order to calculate the ExFT g-functions.

Author Contributions: Both authors contributed to this work as follows: Conceptualization, M.S.M. and J.D.S.; methodology, M.S.M. and J.D.S.; software, M.S.M; validation, M.S.M. and J.D.S.; formal analysis, M.S.M. and J.D.S.; investigation, M.S.M. and J.D.S.; resources, M.S.M. and J.D.S.; data curation, M.S.M. and J.D.S.; writing—original draft preparation, M.S.M. and J.D.S.; writing—review and editing, M.S.M. and J.D.S.; visualization, M.S.M. and J.D.S.; supervision, J.D.S.; project administration, M.S.M.; funding acquisition, M.S.M. All authors have read and agreed to the published version of the manuscript.

Funding: This work was authored in part by the National Renewable Energy Laboratory, operated by Alliance for Sustainable Energy, LLC, for the U.S. Department of Energy (DOE) under Contract No. DE-AC36-08GO28308. Funding was provided by the U.S. Department of Energy Office of Energy Efficiency and Renewable Energy Building Technologies Office. The views expressed in the article do not necessarily represent the views of the DOE or the U.S. Government. The U.S. Government retains and the publisher, by accepting the article for publication, acknowledges that the U.S. Government retains a nonexclusive, paid-up, irrevocable, worldwide license to publish or reproduce the published form of this work, or allow others to do so, for U.S. Government purposes. The work performed was partially supported under the National Renewable Energy Laboratory Task Order Agreement KAGN-4-42503-00.

Acknowledgments: The authors gratefully acknowledge Oklahoma State University and The National Renewable Energy Laboratory for supporting this work.

Conflicts of Interest: The authors declare no conflict of interest.

Abbreviations

The following abbreviations are used in this manuscript:

ExFT	exiting fluid temperature
FLS	finite line source
GHE	ground heat exchanger
GSHP	ground-source heat pump
ICS	infinite cylinder source
ILS	infinite line source
MFRTRT	multi-flowrate thermal response test
TRC	thermal resistance-capacitance
TRT	thermal response test
WBES	whole-building energy simulation

References

1. Crawley, D.B.; Lawrie, L.K.; Winkelmann, F.C.; Buhl, W.F.; Huang, Y.J.; Pedersen, C.O.; Strand, R.K.; Liesen, R.J.; Fisher, D.E.; Witte, M.J.; et al. EnergyPlus: Creating a new-generation building energy simulation program. *Energy Build.* **2001**, *33*, 319–331. [CrossRef]
2. Klein, S.A.; Beckman, W.A.; Mitchell, J.W.; Duffie, J.A.; Duffie, N.A.; Freeman, T.; Mitchell, J.C.; Braun, J.E.; Evans, B.L.; Kummer, J.P.; et al. *TRNSYS Manual: A Transient Simulation Program*; Solar Engineering Laboratory; University of Wisconsin-Madison: Madison, WI, USA, 2016.
3. Li, M.; Zhu, K.; Fang, Z. Analytical methods for thermal analysis of vertical ground heat exchangers. In *Advances in Ground-Source Heat Pump Systems*; Woodhead Publishing: Duxford, UK, 2016; pp. 157–185.
4. Spitler, J.D.; Bernier, M.A. Vertical borehole ground heat exchanger design methods. In *Advances in Ground-Source Heat Pump Systems*; Woodhead Publishing: Duxford, UK, 2016; pp. 29–61.
5. Thomson, W. Compendium of the Fourier Mathematics for the Conduction of Heat in Solids, and the Mathematically Allied Physical Subjects of Diffusion of Fluids, and Transmission of Electrical Signals Through Submarine Cables. In *Article 72. Mathematical and Physical Papers*; University Press: Cambridge, UK, 1884; pp. 41–60.
6. Spitler, J.D.; Gehlin, S.E.A. Thermal response testing for ground source heat pump systems—An historical review. *Renew. Sustain. Energy Rev.* **2015**, *50*, 1125–1137. [CrossRef]
7. Carslaw, H.S.; Jaeger, J.C. *Conduction of Heat in Solids*, 2nd ed.; Clarendon Press: London, UK, 1959.
8. Hellström, G. Ground Heat Storage: Thermal Analysis of Duct Storage Systems. Ph.D. Thesis, University of Lund, Lund, Sweden, 1991.
9. Mogensen, P. Fluid to duct wall heat transfer in dust system heat storages. In *International Conference on Subsurface Heat Storage in Theory and Practice*; Swedish Council for Building Research: Stockholm, Sweden, 1983.
10. Bernier, M.A. Ground-coupled heat pump system simulation. *ASHRAE Trans.* **2001**, *107*, 605–616.
11. Ingersoll, L.R.; Zobel, O.J.; Ingersoll, A.C. *Heat Conduction: With Engineering and Geological Applications*, 2nd ed.; McGraw-Hill: New York, NY, USA, 1954.
12. Diao, N.R.; Li, Q.Y.; Fang, Z. Heat transfer in ground heat exchangers with ground water advection. *Int. J. Therm. Sci.* **2004**, *43*, 1203–1211. [CrossRef]
13. Man, Y.; Yang, H.; Diao, N.R.; Liu, J.; Fang, Z.H. A new model and analytical solutions for borehole and pile ground heat exchangers. *Int. J. Heat Mass Transf.* **2010**, *53*, 2593–2601. [CrossRef]
14. Li, M.; Lai, L.C.K. Heat-source solutions to heat conduction in anisotropic media with applications to pile and borehole ground heat exchangers. *Appl. Energy* **2012**, *96*, 451–458. [CrossRef]
15. Marcotte, D.; Pasquier, P. The effect of borehole inclination on fluid and ground temperature for GLHE systems. *Geothermics* **2009**, *38*, 392–398. [CrossRef]
16. Lamarche, L.; Beauchamp, B. A new contribution to the finite line-source model for geothermal boreholes. *Energy Build.* **2007**, *39*, 188–198. [CrossRef]
17. Javed, S.; Claesson, J. New analytical and numerical solutions for the short-term analysis of vertical ground heat exchangers. *ASHRAE Trans.* **2011**, *117*, 3–12.

18. Grundmann, R.M. Improved Design Methods for Ground Heat Exchangers. Master's Thesis, Oklahoma State University, Stillwater, OK, USA, 2016.

19. Spitler, J.D. GLHEPro—A design tool for commercial building ground loop heat exchangers. In Proceedings of the 4th International Heat Pump in Cold Climates Conference, Alymer, QC, Canada, 17–18 August 2000.

20. Duhamel, M.J.M.C. Sur les equations générales de la propagation de la chaleur dan les corps solides dont la conductiblité n'est pas la même dan tous les sens. *J. L'Ecole Polytech.* **1828**, *21*, 356–399.

21. Özişik, M.N. *Boundary Value Problems of Heat Conduction*; Dover Publications: New York, NY, USA, 2002.

22. Claesson, J.; Dunand, A. *Heat Extraction from the Ground by Horizontal Pipes—A Mathematical Analysis*; Technical Report; Swedish Council for Building Research: Stockholm, Sweden, 1983.

23. Eskilson, P. Thermal Analysis of Heat Extraction Boreholes. Ph.D. Thesis, University of Lund, Lund, Sweden, 1987.

24. Claesson, J.; Eskilson, P. Conductive heat extraction to a deep borehole—Thermal analysis and dimensioning rules. *Energy* **1988**, *13*, 509–527. [CrossRef]

25. Eskilson, P.; Claesson, J. Simulation model for thermally interacting heat extraction boreholes. *Numer. Heat Transf.* **1988**, *13*, 149–165. [CrossRef]

26. Mitchell, M.S.; Spitler, J.D. Characterization, testing, and optimization of load aggregation methods for ground heat exchanger response-factor models. *Sci. Technol. Built Environ.* **2019**, *8*, 1036–1051. [CrossRef]

27. Brussieux, Y.; Bernier, M. Universal short time g*-functions: generation and application. *Sci. Technol. Built Environ.* **2019**, *8*, 993–1006. [CrossRef]

28. Loveridge, F.; Powrie, W. Temperature response functions (G-functions) for single pile heat exchangers. *Energy* **2013**, *57*, 554–564. [CrossRef]

29. Alberdi-Pagola, M. Design and Performance of Energy Pile Foundations. Ph.D. Thesis, Aalborg University, Aalborg, Denmark, 2018.

30. Alberdi-Pagola, M.; Jensen, R.L.; Poulsen, S.; Erbs, S. *Method to Obtain G-Functions for Multiple Precast Quadratic Pile Heat Exchangers*; Technical Report 243; Aalborg University: Aalborg Øst, Denmark, 2018.

31. Dusseault, B.; Pasquier, P. Efficient g-function approximation with artificial neural networks for a varying number of boreholes on a regular or irregular layout. *Sci. Technol. Built Environ.* **2019**, *8*, 1023–1035. [CrossRef]

32. Pasquier, P.; Zarrella, A.; Labib, R. Application of artificial neural networks to near-instant construction of short-term g-functions. *Appl. Therm. Eng.* **2018**, *143*, 910–921. [CrossRef]

33. Pasquier, P.; Marcotte, D. Short-term simulation of ground heat exchanger with an improved TRCM. *Renew. Energy* **2012**, *46*, 92–99. [CrossRef]

34. Pasquier, P.; Marcotte, D. Joint use of quasi-3D response model and spectral method to simulate borehole heat exchanger. *Geothermics* **2014**, *51*, 281–299. [CrossRef]

35. Cimmino, M.; Bernier, M. A semi-analytical method to generate g-functions for geothermal bore fields. *Int. J. Heat Mass Transf.* **2014**, *70*, 641–650. [CrossRef]

36. Malayappan, V.; Spitler, J.D. Limitation of using uniform heat flux assumptions in sizing vertical borehole heat exchanger fields. In Proceedings of the CLIMA 2013, Prague, Czech Republic, 16–19 June 2013.

37. Marcotte, D.; Pasquier, P. Unit-response function for ground heat exchanger with parallel, series or mixed borehole arrangement. *Renew. Energy* **2014**, *68*, 14–24. [CrossRef]

38. Claesson, J.; Javed, S. An analytical method to compute borehole fluid temperatures for timescales from minutes to decades. *ASHRAE Trans.* **2011**, *117*, 279–288.

39. Cimmino, M. Fast calculation of the g-functions of geothermal borehole fields using similarities in the evaluation of the finite line source solution. *J. Build. Perform. Simul.* **2018**, *11*, 655–668. [CrossRef]

40. Cimmino, M. g-Functions for bore fields with mixed parallel and series connections considering axial fluid temperature variations. In Proceedings of the IGSHPA Research Conference Proceedings, IGSHPA, Stockholm, Sweden, 18–19 September 2018; pp. 262–271.

41. Cimmino, M. Semi-analytical method for g-function calculation of bore fields with series- and parallel-connected boreholes. *Sci. Technol. Built Environ.* **2019**, *25*, 1007–1022. [CrossRef]

42. Cimmino, M. pygfunction: An open-source toolbox for the evaluation of thermal response factors for geothermal borehole fields. In Proceedings of the eSim 2018, the 10th Conference of IBPSA-Canada, International Building Performance Simulation Association, Montréal, QC, Canada, 9 March 2018; pp. 492–501. [CrossRef]

43. De Carli, M.; Tonon, M.; Zarrella, A.; Zecchin, R. A computational capacity resistance model (CaRM) for vertical ground-coupled heat exchangers. *Renew. Energy* **2010**, *35*, 1537–1550. [CrossRef]
44. Zarrella, A.; Scarpa, M.; De Carli, M. Short time step analysis of vertical ground-coupled heat exchangers: the approach of CaRM. *Renew. Energy* **2011**, *36*, 2357–2367. [CrossRef]
45. Bauer, D.; Heidemann, W.; Müller-Steinhagen, H.; Diersch, H.J.G. Thermal resistance and capacity models for borehole heat exchangers. *Int. J. Energy Res.* **2011**, *35*, 312–320. [CrossRef]
46. Ruiz-Calvo, F.; De Rosa, M.; Acuña, J.; Corberán, J.M.; Montagud, C. Experimental validation of a short-term borehole-to-ground (B2G) dynamic model. *Appl. Energy* **2015**, *140*, 210–223. [CrossRef]
47. Cazorla-Marín, A.; Montagud, C.; Montero, Á.; Martos, J.; Corberán, J.M. Influence of different ground thermal properties in a borehole heat exchanger's performance using the B2G dynamic model. In Proceedings of the IGSHPA Research Conference, IGSHPA, Denver, CO, USA, 14–16 March 2017; pp. 192–200.
48. Ruiz-Calvo, F.; De Rosa, M.; Monzó, P.; Montagud, C.; Corberán, J.M. Coupling short-term (B2G model) and long-term (g-function) models for ground source heat exchanger simulation in TRNSYS. Application in a real installation. *Appl. Therm. Eng.* **2016**, *102*, 720–732. [CrossRef]
49. Al-Khoury, R. Spectral framework for geothermal borehole heat exchangers. *Int. J. Numer. Methods Heat Fluid Flow* **2010**, *20*, 773–793. [CrossRef]
50. Shao, H.; Hein, P.; Sachse, A.; Kolditz, O. *Geoenergy Modeling II: Shallow Geothermal Systems*; Springer: Cham, Switzerland, 2016.
51. Al-Khoury, R.; Bonnier, P.G.; Brinkgreve, R.B.J. Efficient finite element formulation for geothermal heating systems–Part I: Steady state. *Int. J. Numer. Methods Eng.* **2005**, *63*, 988–1013. [CrossRef]
52. Al-Khoury, R.; Bonnier, P.G. Efficient finite element formulation for geothermal heating systems–Part II: Transient. *Int. J. Numer. Methods Eng.* **2006**, *67*, 725–745. [CrossRef]
53. He, M.; Rees, S.; Shao, L. Simulation of a domestic ground source heat pump system using a three-dimensional numerical borehole heat exchanger model. *J. Build. Perform. Simul.* **2011**, *4*, 141–155. [CrossRef]
54. Nabi, M.; Al-Khoury, R. An efficient finite volume model for shallow geothermal systems–Part I: Model formulation. *Comput. Geosci.* **2012**, *49*, 290–296. [CrossRef]
55. Nabi, M.; Al-Khoury, R. An efficient finite volume model for shallow geothermal systems–Part II: Verification, validation and grid convergence. *Comput. Geosci.* **2012**, *49*, 297–307. [CrossRef]
56. Fisher, D.E.; Rees, S.J.; Padhmanabhan, S.K.; Murugappan, A. Implementation and validation of ground-source heat pump system models in an integrated building and system simulation environment. *HVAC&R Res.* **2006**, *12*, 693–710. [CrossRef]
57. Yavuzturk, C.; Spitler, J.D. A short time step response factor model for vertical ground loop heat exchangers. *ASHRAE Trans.* **1999**, *105*, 475–485.
58. Xu, X.; Spitler, J.D. Modeling of vertical ground loop heat exchangers with variable convective resistance and thermal mass of the fluid. In Proceedings of the 10th International Conference on Thermal Energy Storage-EcoStock, Pomona, NJ, USA, 31 May–2 June 2006.
59. Mitchell, M.S. An Enhanced Vertical Ground Heat Exchanger Model for Whole Building Energy Simulation. Ph.D. Thesis, Oklahoma State University, Stillwater, OK, USA, 2019.
60. Loveridge, F. The Thermal Performance of Foundation Piles Used as Heat Exchangers in Ground Energy Systems. Ph.D. Thesis, University of Southampton, Southampton, UK, 2012.
61. Bennet, J.; Claesson, J.; Helström, G. *Multipole Method to Compute the Conductive Heat Flow to and between Pipes in a Composite Cylinder*; Technical Report; University of Lund, Department of Building and Mathematical Physics: Lund, Sweden, 1987.
62. Claesson, J. *Multipole Method to Calculate Borehole Thermal Resistances: Mathematical Report*; Technical Report 2012-06-20; Chalmers Technical University: Gothenburg, Sweden, 2011.
63. Claesson, J.; Hellström, G. Multipole method to calculate borehole thermal resistances in a borehole heat exchanger. *HVAC&R Res.* **2011**, *17*, 895–911. [CrossRef]
64. Javed, S.; Spitler, J.D. Accuracy of borehole thermal resistance calculation methods for grouted single u-tube ground heat exchangers. *Appl. Energy* **2017**, *187*, 790–806. [CrossRef]
65. Javed, S.; Spitler, J.D. Chapter Calculation of Borehole Thermal Resistance. In *Advances in Ground-Source Heat Pump Systems*; Woodhead Publishing: Duxford, UK, 2016; pp. 63–95.
66. Moin, P. *Fundamentals of Engineering Numerical Analysis*; Cambridge University Press: New York, NY, USA, 2010.

67. Bischoff, K.B.; Levenspiel, O. Fluid dispersion—Generalization and comparison of mathematical models—II comparison of models. *Chem. Eng. Sci.* **1962**, *17*, 257–264. [CrossRef]
68. Skoglund, T.; Dejmek, P. A dynamic object-oriented model for efficient simulation of fluid dispersion in turbulent flow with varying fluid properties. *Chem. Eng. Sci.* **2007**, *62*, 2168–2178. [CrossRef]
69. Rees, S.J. An extended two-dimensional borehole heat exchanger model for simulation of short and medium timescale thermal response. *Renew. Energy* **2015**, *83*, 518–526. [CrossRef]
70. Beier, R.A.; Mitchell, M.S.; Spitler, J.D.; Javed, S. Validation of borehole heat exchanger models against multi-flow rate thermal response tests. *Geothermics* **2018**, *71*, 55–68. [CrossRef]
71. Woodruff, M.J.; Herman, J. Pareto. 2018. Available online: https://github.com/matthewjwoodruff/pareto.py (accessed on 1 October 2018).
72. Deb, K.; Pratap, A.; Agarwal, S.; Meyarivan, T. A fast and elitist multiobjective genetic algorithm: NSGA-II. *IEEE Trans. Evol. Comput.* **2002**, *6*, 182–197. [CrossRef]
73. Bell, I.H.; Wronski, J.; Quoilin, S.; Lemort, V. Pure and pseudo-pure fluid thermophysical property evaluation and the open-source thermophysical property library CoolProp. *Ind. Eng. Chem. Res.* **2014**, *53*, 2498–2508. [CrossRef] [PubMed]
74. Patankar, S.V. *Computation of Conduction and Duct Flow Heat Transfer*; Innovative Research, Inc.: Maple Grove, MN, USA, 1991.

Article

Experimental Hydration Temperature Increase in Borehole Heat Exchangers during Thermal Response Tests for Geothermal Heat Pump Design

Fabio Minchio [1,*], Gabriele Cesari [2], Claudio Pastore [2] and Marco Fossa [3]

[1] Studio 3F Engineering, Via IV Novembre 14, 36051 Creazzo, Italy
[2] Geo-Net S.r.l., Via Giuseppe Saragat 5, 40026 Imola, Italy; g.cesari@geo-net.it (G.C.);
     geologia@geo-net.it (C.P.)
[3] DIME Department of Mechanical, Energy, Management and Transportation Engineering,
     the University of Genova, Via Opera Pia 15a, 16145 Genova, Italy; marco.fossa@unige.it
*   Correspondence: fabio.minchio@gmail.com

Received: 7 June 2020; Accepted: 30 June 2020; Published: 4 July 2020

Abstract: The correct design of a system of borehole heat exchangers (BHEs) is the primary requirement for attaining high performance with geothermal heat pumps. The design procedure is based on a reliable estimate of ground thermal properties, which can be assessed by a Thermal Response Test (TRT). The TRT analysis is usually performed adopting the Infinite Line Source model and is based on a series of assumptions to which the experiment must comply, including stable initial ground temperatures and a constant heat transfer rate during the experiment. The present paper novelty is related to depth distributed temperature measurements in a series of TRT experiments. The approach is based on the use of special submersible sensors able to record their position inside the pipes. The focus is on the early period of BHE installation, when the grout cement filling the BHE is still chemically reacting, thus releasing extra heat. The comprehensive dataset presented here shows how grout hydration can affect the depth profile of the undisturbed ground temperature and how the temperature evolution in time and space can be used for assessing the correct recovery period for starting the TRT experiment and inferring information on grouting defects along the BHE depth.

Keywords: ground coupled heat pumps; borehole heat exchangers; distributed temperature response test; grouting material; hydration heat release

---

1. Introduction

Nowadays, geothermal heat exchanger facilities are in rapid growth and the research for new instrumentation and surveying methods is in constant evolution in the direction of reliable and low-cost systems.

Borehole heat exchangers (BHEs) for ground coupled heat pump (GCHP) applications are the most popular technology in low enthalpy applications, where heat stored in the ground is exploited for heating purposes in building applications. In this context the knowledge of the geological, hydrogeological, and thermal conditions of the ground medium is fundamental for the system efficiency and economical sustainability. Ground thermal conductivity is one of these key parameters, since it controls the heat transfer rate between the BHE field and the ground mass [1–3].

In addition to favorable thermal conductivity and specific heat, the presence of groundwater and the related advection phenomena can further increase the heat transfer performance of the BHE field [4,5].

The International Energy Agency (IEA ECES) [6] and the Ente Nazionale di Unificazione (UNI) [7] standards recognize the Thermal Response Test (TRT) as the best practice in order to estimate the ground thermal properties and from them to calculate the optimal length and distribution of BHEs. As it is well known, the TRT method allows the effective ground thermal conductivity (k_{gr}) and effective borehole thermal resistance (R_{bhe}) to be estimated from temperature measurements at the BHE top inlet and outlet ports.

However, the standard TRT experiment is not able to provide detailed information on ground properties, since the test is not conceived for retrieving information about the heterogeneity of ground lithology and the presence of aquifers, thus leading to a bad long-term system performance and wrong BHE field design [1,8]. To define the optimum borefield geometry for the best trade-off between energy performance and investment costs (e.g., drilling and grouting), additional information on ground property distribution along the BHE depth are necessary [9].

To address this lack of information, several tools have been developed in order to evaluate the depth-related ground properties along the BHE. For instance, a device developed by Martos et al. [10] called Geoball consists of a wireless sensor with a spherical shape and embedded data logger temperature and position. The ball-shaped sensor has a density similar to that of the fluid, thus it is able to be carried by a flow inside the BHE pipe. The exact sensor location along the pipes can be calculated from pressure measurements during the sink of the sensor itself.

Commercial companies have started to realize similar devices. Among them, the enOware enterprise developed its integrated sensor (named GEOsniff, [11]), a ball-shaped device able to record temperature and position while travelling along the pipe of a pilot BHE. This submersible sensor has a higher density than water, allowing it to sink along the pipes. To expel the sensor from the pipe, it has to be flushed out in a tank using a pump.

Haranzabal et al. [1,12] have tested and compared different sensors for distributed (along the BHE depth) measurements, including submersible sensors and optical fibers, confirming that travelling sensors inside the carrier fluid stream are suitable for temperature measurements along the BHE. Furthermore, according to the literature investigation [1,12] these sensors can assure good spatial accuracy, provided that the sinking speed and sampling time are correctly chosen.

Once the temperature values have been made available by such distributed measurements along the BHE depth, ground properties and even grouting characteristic estimations can be achieved with theoretical models able to describe the heat transfer phenomena along the BHE radius [13–15].

The importance of a correct grouting during borehole heat exchangers installation is well known [16–18]. The grout is used to seal and stabilize the borehole, providing a heat transfer medium between pipes and the surrounding ground. In addition, the grout is able to act as a barrier against groundwater movement and contamination in the vertical direction.

The grout affects clearly the effective borehole thermal resistance R_{bhe}, as discussed in several studies [13–15,19–21]. Zeng et al. [15] in particular obtained a complete set of equations for R_{bhe} calculation and demonstrated that the U-tube shank spacing and grout thermal conductivity are the prevailing factors in determining the R_{bhe}.

It is hence important to select the proper grouting material, optimizing the thermal conductivity with respect to the ground thermal one and considering also other grout properties such as the coefficient of permeability in the bulk state and, when bonded to BHE tubes, shrinkage, bond strength, wet-dry and freeze-thaw durability, leachability, exotherm, and the coefficient of thermal expansion. Since the late 1990s, research institutes [22] and ground heat pump companies have developed different types of enhanced thermal grouts, the use of which is recommended in most design guidelines.

Hossein et al. [23] provide a comprehensive review of backfill materials in geothermal applications. The six most common grout types used in geothermal installations are pure bentonite, neat cement, cement-bentonite grouts, bentonite/sand mixtures, cementitious grouts containing fillers (such as PFA, Pulverized Fuel Ash [24], or micro-fine colloidal silica and colloidal lime or others), and graphite-based materials [25]. There is also recent development in the addition of phase change materials (PCM) [26].

It is also very important to properly backfill the borehole during the installation, as the IGHSPA Association recommends in its specific grouting dedicated manual [16]. Improper grouting has a negative impact on the heat transfer process, modifying the thermal behavior of the geothermal field with respect to the design conditions, resulting in a worse energy performance of the geothermal heat pump system and, in severe cases, critical system operation problems.

Philippacopoulos and Berndt [27] investigated the influence of debonding in ground heat exchangers. They found that debonding at the backfill/pipe interface has greater significance than debonding between grout and surrounding formation. Chen and Mao [28] studied the influence of grout backfilling, estimating the thermal resistance through a 2D steady-state heat conduction model in four different cases: (1) fine backfilling, (2) porous backfilling, (3) hole-wall delaminating, and (4) pipe-wall delaminating. The comparison of performance shows that the heat transfer rate per unit length at the borehole wall decreases by 6.3%, 41.5%, and 78.4%, respectively, in the case of porous backfilling, hole-wall delaminating, and pipe-wall delaminating, with pipe-wall detaching having the worst impact on the BHE performance.

It is quite difficult to find a way to check and control the correct BHE grouting. Considering that most grouting materials contain cement, it is possible to use the hydration reaction to detect the quality of grouting by in situ measurements because the grouting hardening reaction (hydration) is exothermic. The hydration temperature profile depends on the type of grout, the water content, the presence of air or other materials, and the thermal transport potential of the surrounding ground, as Suibert Oskar Seibertz et al. [29] discussed based on their field and laboratory tests.

If a simple and quick way to monitor the heat of hydration effects could be found, it would be possible to complete the BHE test installation with a grouting quality investigation. Thus, devices such as submergible sensors could be a good solution for this purpose too.

The present work is aimed at developing an experimental technique based on distributed temperature measurements along the depth of pilot BHEs in order to exploit the exothermic hardening reaction of grout for assessing the presence of thermal anomalies along the heat exchanger depth. The measurements refer to a large series of real BHE installations drilled in Italy, and in this sense the present investigation represents the very first one on a large scale devoted to the thermal analysis of grout material in the early stage of BHE installation. It is here demonstrated that grouting chemical heating can be exploited by distributed temperature measurements for precisely detecting ground anisotropies at vertical layers and grouting defects. An additional series of experiments is here discussed for assessing the influence of the hydration heating on TRT experiments, including the estimation of the time needed for assuring a reliable estimation of the ground undisturbed temperature.

2. Thermal Response Test Theory

Ground thermal properties are typically inferred during a TRT experiment, during which a constant flow rate of water is circulated in a pilot BHE while a constant heat transfer rate is supplied to the carrier fluid by some electric heater inside the TRT machine, which is located at the ground surface.

The measurements of the inlet and outlet fluid temperatures at the pilot BHE allow the ground thermal conductivity to be estimated based on the heat transfer rate and undisturbed ground temperature knowledge, both estimated during the experiment itself. The method was first proposed by Palne Mogensen [30], a Swedish Engineer who first realized that the Infinite Line Source Model (ILS), usually applied to small samples at laboratory level, could be applied to the large volumes of ground surrounding a borehole heat exchanger. Mogensen conducted his own TRT experiments with a refrigeration machine able to extract heat from the carrier fluid at an almost constant rate. Mogensen also realized that the ILS model and its log linear approximation could be applied for describing the fluid temperature evolution in time during the TRT experiment, provided that an extra term to the ILS expression was added ("the thermal transfer resistance" in the Mogensen paper, later by other researchers "the effective BHE resistance", R_{bhe}).

After Mogensen, a series of researchers applied the method and realized dedicated equipment. Among them, one can refer to Gehlin [31], who first realized the first mobile TRT machine; Austin [32], who first performed TRT experiments in the US; Acuna and co-workers [33], who first employed distributed sensors along the BHE in order to infer the ground thermal properties along the ground depth; and Fossa et al. [34], who described the possibility of working at a non-constant TRT heat transfer rate and the advantages in terms of the property estimates of such pulsated experiments. Very recently, Morchio and Fossa [35] tackled the problem of TRT analysis with deep boreholes (up to 800 m), when the geothermal gradient effects can lead to temperature crossing in the bottom part of the heat exchanger. From the same research group, the TRT analysis in terms of proper temperature response factors is further extended to geothermal pipes [36]. Franco and Conti [37] present an updated and comprehensive review of the various TRT types and explore the perspective to combine TRT and routine geotechnical tests.

The Infinite Line Source (ILS) was first described with reference to ground heat exchanger problems by Ingersoll et al. [38] in their book. The ILS approach provides an analytical solution of the ground's temperature evolution in time and radial space when a constant heat transfer rate (per unit length $\dot{Q}'$) is applied to infinitely long linear sources buried inside an infinite medium. In dimensionless form, the ground temperature T can be expressed as an excess with respect to the undisturbed value $T_{gr,\infty}$:

$$\frac{T(r,\tau) - T_{gr,\infty}}{\dot{Q}'/2\pi k_{gr}} = \frac{1}{2} \int_{1/4Fo_r}^{\infty} \frac{e^{-\beta}}{\beta} d\beta = \frac{1}{2} E_1(1/4Fo_r), \tag{1}$$

where k_{gr} is the ground thermal conductivity and Fo_r is the Fourier number based on the radial distance from the line source, r. In the above expression, the complex Exponential Integral E_1 can be accurately approximated by simple series expansions, as discussed, for example, by Fossa [39]. Back to the Mogensen intuition, once the interest is on the fluid temperature and the fluid is circulating inside a BHE pipe which locates some r_b (borehole radius) from the ground medium (where ILS is expected to apply), additional thermal resistance must be added to the ground thermal resistance R_{gr}:

$$R_{gr} = E_1(Fo_{rb})/4\pi k_{gr}. \tag{2}$$

This additional thermal resistance is the above-mentioned borehole effective resistance R_{bhe} (after Eskilson definition, [40]). Moreover, one can adopt the second term truncated version of the E_1 expansion series, provided that the Fo_{rb} range of interest is higher than about 10 [38]:

$$E_1(Fo_{rb}) \cong -\gamma - \ln(1/4Fo_{rb}), \tag{3}$$

where γ is the Euler constant. The final step is to rearrange the two thermal resistance model in a proper way in order to allow the k_{gr} estimation:

$$T_{f,ave}(\tau) - T_{gr,\infty} = \dot{Q}'\left[R_{bhe} + \frac{1}{4\pi\,k_{gr}}\left(-\gamma - \ln\left(\frac{1}{4Fo_{rb}}\right)\right)\right]. \tag{4}$$

Equation (4) can be easily thought as a linear expression of the fluid temperature (as the inlet/outlet average outside the BHE) as a function of the logarithm of time. The new expression contains the line slope m and a constant, which in turn is a function of R_{bhe}:

$$T_{f,ave} = a + m\,\ln(\tau). \tag{5}$$

Finally, the ground thermal conductivity can be calculated from the estimate of slope m according to the expression:

$$k_{gr} = \frac{\dot{Q}'}{4\,\pi\,m} \tag{6}$$

For solving the above equations system, the undisturbed (initial) ground temperature $T_{gr,\infty}$ is required, and it is usually inferred during the first part of the TRT experiment, when the carrier fluid is circulated without any heat injection or extraction (adiabatic part of the test).

The borehole thermal resistance R_{bhe} is itself chained with k_{gr}, and its estimation is possible for any instantaneous fluid temperature value, such as:

$$R_{bhe}(\tau) = \frac{T_{f,ave} - T_{gr,\infty}}{\dot{Q}'} - \frac{1}{4\pi \, k_{gr}}\left(-\gamma - \ln\left(\frac{1}{4Fo_{rb}}\right)\right). \tag{7}$$

It is apparent from the above set of equations that the correct estimation of both k_{gr} and R_{bhe} is strictly related to the knowledge of the undisturbed ground temperature, which in turn is measured though the fluid temperature during the initial adiabatic part of the TRT run, hence provided that any heat source is present at this stage of the experiment.

3. Grouting and Hydration Reaction

The hydration of Portland cement refers to a series of chemical reactions taking place within a water–cement system: the silicates and alumina react with water to form hydration products. There are four major mineral compounds in Portland cement: C3S, C2S, C3A, and C4AF, which are hydration exothermic reactions which differ significantly from each other.

The heat generated by the cement's hydration raises the temperature of the grout material inside the BHE. An example of the rate of heat development versus hydration time curve is illustrated in Figure 1 [41], where it is possible to note the typical five stages of hydration reaction: the initial reaction, the induction period, the acceleration period, the deceleration period, and the steady state condition.

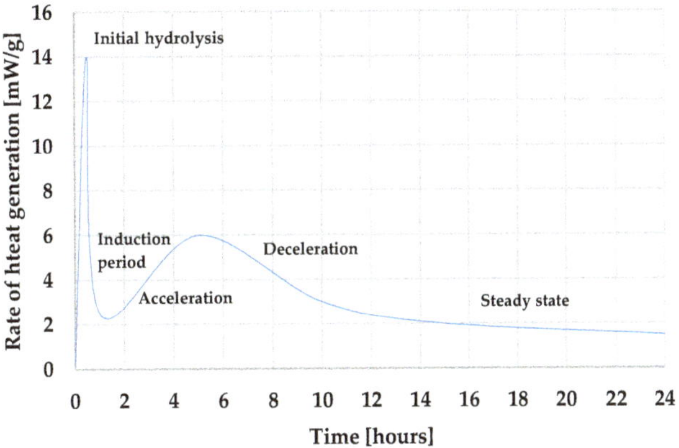

Figure 1. An example of the rate of heat development versus hydration time curve.

The heat of hydration influences the temperatures inside and around the BHE during the first days after installation, depending on the grout mixture properties and chemical composition and in particular its cement percentage. The time required for the ground to return to an approximately undisturbed state after the installation was not enough systematically investigate. Kavanaugh [42] recommends to wait before performing a TRT for at least 24 h after drilling, and at least 72 h if cementitious grouts are used. The ASHRAE Fundamentals Handbook [43] TRT guidelines prescribes the following: "A waiting period of five days is suggested for low-conductivity soils [k < 1.7 W/(m·K)] after the ground loop has been installed and grouted (or filled) before the thermal conductivity test is initiated. A delay of three days is recommended for higher-conductivity formations [k > 1.7 W/(m K)]." The Italian standard UNI 11466 [7] at nomenclature, reports identical recommendations.

With distributed temperature measurements performed immediately and during the first hours after BHE installation and grouting (i.e., during the hardening process), it is thus possible to observe how the hydration reaction affects the ground temperature profile and therefore investigate the quality of the grouting itself. In fact, in the presence of a small rise in temperature over time at a specific ground layer with respect to other layers, it is thus possible to guess there is poor grouting at that position.

It is worth noticing that the hydration reaction does not affect the medium and long term performance of the BHE and the coupled geothermal system, since the hydration heat input inside the ground is very small (compared to GCHP operations) and very short in time (a few days). It is, however, important to stress that any TRT must not to be performed until the hydration reaction has ended, and to this aim the knowledge of the hydration phenomenon duration is fundamental. Furthermore, a thermal analysis like the present one during the hardening process can reveal poor grouting at specific ground layers, a condition which in turn yields higher values of R_{bhe} with respect to the design values, thus worsening the long-term performance of the whole system.

4. Experimental Apparatus and Test Sites

In TRT experiments, the classic instrumentation employs electric heaters, a circulation pump, and proper sensors (e.g., RTD ones) for measuring the carried fluid temperature at the ground surface [34].

The present investigation applies a different approach to TRT experiments which is based on the use of small submersible sensors, able to travel inside the carrier fluid along the BHE pipes while recording both the temperature and pressure for a hydrostatic estimation of the sensor position along the BHE depth. To heat the fluid and its surrounding ground by 20 W/m, a heating cable (5 mm external diameter) is here inserted inside the pipe. This cable is able to deliver a constant heat flux along the pipe length, and hence it can fulfill the basic assumption behind the Infinite Line Source model.

The heating cable is inserted along one branch of the U pipe, while the temperature measurements are recorded by the travelling sensor in another pipe leg. This method allows us to perform a distributed TRT, since the local measurements allow the depth-related ground properties to be estimated. The equipment used by Geo-Net is the submersible sensor (Figure 2a) produced by ENOWARE GmbH [11]. The equipment is composed by the so-called Validation Box (Figure 2b), where data is downloaded from the floating sensor and processed for inferring ground related properties.

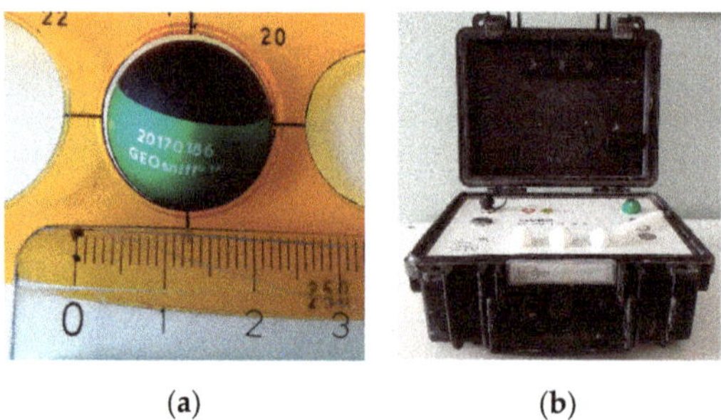

(a)   (b)

Figure 2. (a) Submersible sensor (ruler units are centimeters) and (b) its validation box.

The sensor (Figure 2a) has the shape of a small ball, 2 cm in diameter, equipped with both a thermometer and a pressure meter to estimate the depth position of the sensor when slowly moving inside the stagnant fluid inside the BHE pipe (Figure 3). The accuracy of the sensor and its measuring chain is provided by the manufacturer: the temperature resolution is 0.01 K; it has a 95% accuracy, 0.2 °C temperature, and 1 mbar pressure [11].

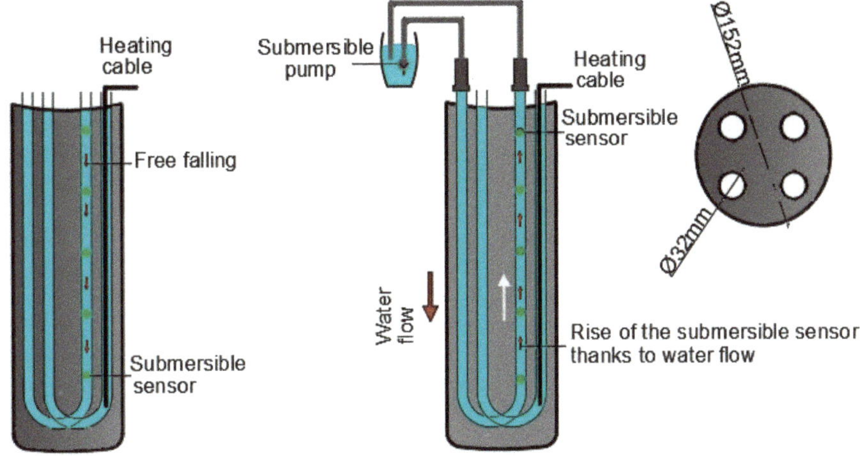

Figure 3. Schematics of the test system, including the heating cable and the submersible sensor moving along one leg of the U-pipe.

The third component of the instrumentation is represented by manufacturer software able to process the data and configure the sensor.

The conversion of pressure into depth preventively requires information about the BHE length and the offset water level within the pipe, together with the density of the fluid (in the present investigation tap water, 998 kg/m^3). The sampling frequency has been set to 1 Hz in order to obtain one measurement every 0.1 m approximately.

The submersible sensor has a known density of 1700 kg/m^3, and the external software is able to estimate the sensor falling time based on the fluid and pipe data.

The dataset collected by the logger is sent to its validation box and then to the software that converts the measurements.

When the temperature measurements are employed for TRT analysis, the Equations (1)–(7) are applied based on the knowledge of the applied heat flux provided by the heating cable. The calculation tool developed by the Authors is able to process measurements pertaining to given depths, thus inferring the ground conductivity k_{gr} and the borehole effective resistance R_{bhe} layer by layer along the BHE depth for proper Fo$_{rb}$ windows.

Test Sites

The case studies presented in this paper refer to a series of real BHE fields located in northern Italy. All the measurements have been performed by the Geo-Net company. In the following, a series of identification codes will be employed.

Codes ASC-1 and ASC-2 refer to experiments carried out in Imola city (Italy, 44°20′29.057″ N; 11°43′23.46″ E). Codes TAS-1 and TAS-2 refer to another pilot BHE in the same city (44°20′50.64″ N; 11°42′16.90″ E). FAE-1 and FAE-2 are BHEs located in Faenza city (44°17′27.06″ N; 11°52′3.85″ E), and finally the PIS code refers to the city of Pistoia (43°57′5.59″ N; 10°53′18.88″ E).

A geological bibliography [44], confirmed by the geological surveys, showed that the ASC-1/2, TAS-1/2, and FAE-1/2 areas belong to the same geological context, which is characterized by continental sedimentary deposits namely made by layers of gravel, sand, and silt. Concerning the PIS site, which is in Tuscany, the geological surveys have shown shale rocks (argillite) surrounded by sedimentary soils consisting of alluvial deposits.

The measurements refer to pilot BHEs constituted by double 32 mm U pipes, and they were performed between 2018 and 2020. All the boreholes are grouted ones, and the BHE nominal diameter is 152 mm. The BHE depth is 120 m both in the ASC and PIS tests, whereas the TAS and FAE experiments were performed with 110 m heat exchangers.

Regarding ASC-2, TAS-1/2, FAE-1/2, and PIS, the measurements refer to the assessment of the vertical ground temperature profiles at different times from grouting the borehole, and they represent a series of snapshots of how the hydration heat can affect the ground undisturbed temperature profile and how such profiles can be employed for an indirect estimation of grouting defects.

Concerning the ASC-1 experiment, a distributed TRT has been performed on one pilot heat exchanger. Such a TRT refers to a series of temperature measurements along the BHE pipe when the electric heater is releasing heat (at a constant rate) inside another pipe of the same BHE.

The adopted procedure was the following one. The TRT had a duration of more than 50 h, according to the recommendations available in [7]. Before activating the heating cable, a temperature measurement is made with the travelling sensor to assess the undisturbed ground temperature as a function of the depth, z (time zero). At this point, the electric cable is powered and a constant in the time heat transfer rate is applied to the BHE. From that instant on, four more records (in the following, referred to as "logs") are carried out after 3, 10, 23, and 50 h from the start of the heating part of the test. The heat transfer rate applied to the heating cable is 2400 W.

5. Results and Discussion

Preliminary measurements with the present experimental technique showed that after some 2 h from the beginning of the BHE grouting, a temperature perturbation can be detected along the BHE depth. This behavior lasts up to 2 or 3 days after the BHE by grout. Figures 4 and 5 show the measured temperature profiles during the grout chemical reaction and after the end of it, some 10 days later. It can be noticed from both figures at the FAE site that the temperature increase (with respect to the undisturbed situation later in time) is almost uniform along the BHE depth, suggesting that the heat release rate per unit depth is uniform. It is worth noticing that, provided that the volumetric heat release is uniform as well, this temperature uniformity can suggest that the grouting was uniform in volume all along the vertical direction.

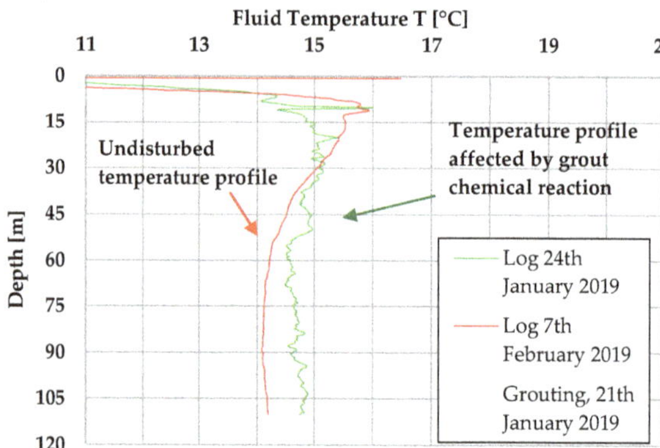

Figure 4. Vertical ground temperature profiles at the FAE-1 pilot borehole heat exchangers (BHE) during the grout hydration period (24 January profile) as compared to one at the end of the chemical reaction, 15 days later.

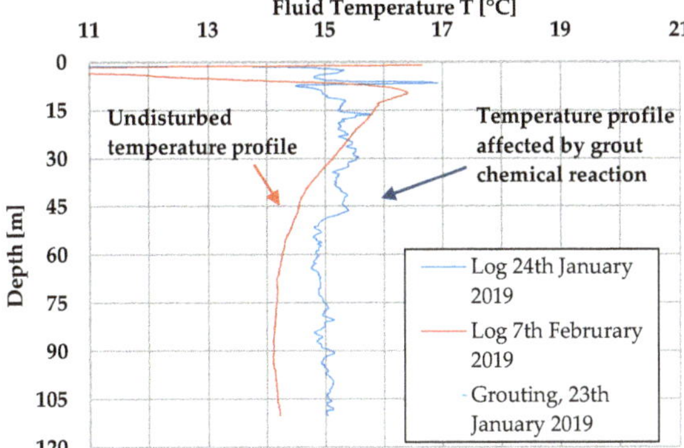

Figure 5. Vertical ground temperature profiles at the FAE-2 pilot BHE during the grout hydration period (24 January profile) as compared to the one at the end of the chemical reaction, 15 days later.

A different situation can be observed from inspecting Figures 6 and 7. Here, the measurements pertain to the PIS and TAS sites, respectively. In particular, Figure 6 shows a zone in the bottom BHE part where the temperature does not change during the grout hydration period. The same can be observed in Figure 7 at the BHE upper section. In both figures, this behavior has been named the "cold zone", which in the experiment described in Figure 6 is related to depths from 55 m on. The presence of cold zones can be observed in Figures 8 and 9 as well in the TAS and ASC measurements.

The presence of cold parts along the BHE depth has to be related to a lower local heat rate or to the absence of it. Furthermore, these low temperatures cannot be ascribed to local heat removal phenomena (e.g., advection), since all the measurements are made inside the grouted volume, which the pipes carrying the electric cable and the travelling sensor are embedded in. As a consequence, the only possible explanation (or better, the more plausible one) is that any cold zone corresponds to a poor grouting, with the presence of air or water pockets.

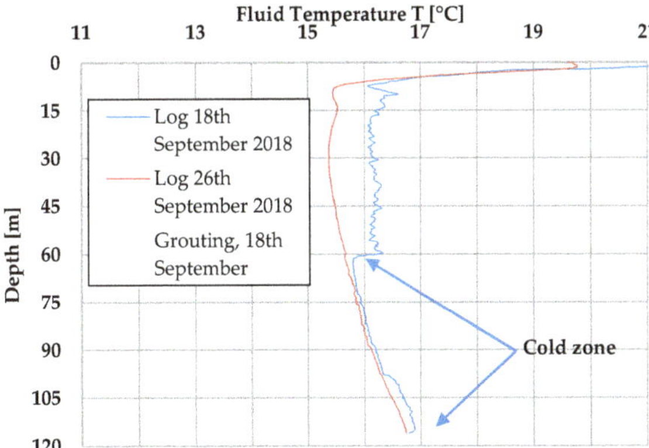

Figure 6. Vertical ground temperature profiles at the PIS pilot BHE during the grout hydration period (18 September profile) as compared to the one at the end of the chemical reaction. A "cold zone" appears at both periods.

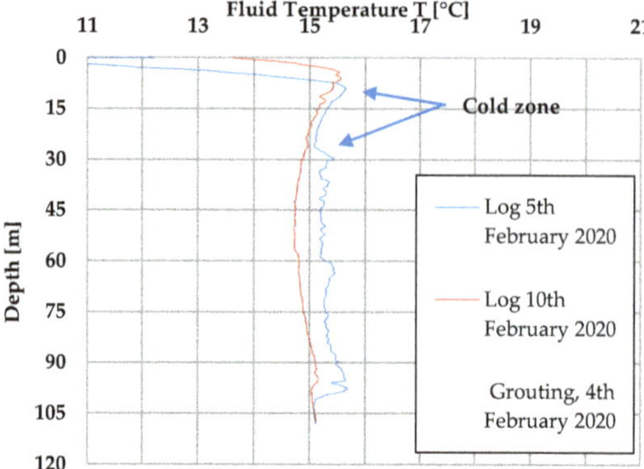

Figure 7. Vertical ground temperature profiles at the TAS-1 pilot BHE during the grout hydration period (5 February profile) as compared to the one at the end of the chemical reaction. A "cold zone" appears at the top BHE.

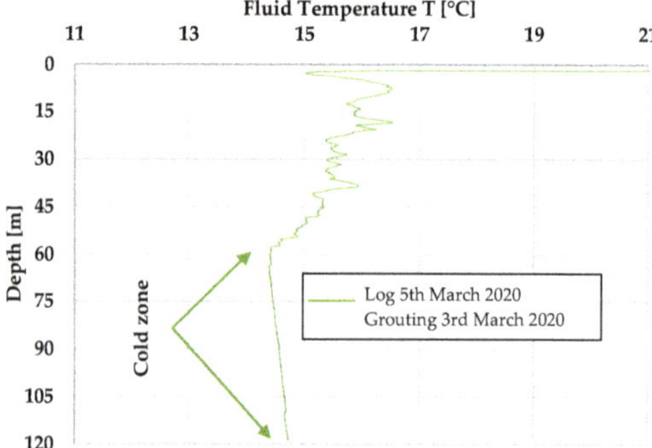

Figure 8. Vertical ground temperature profiles at the ASC-2 BHE during the grout hydration period. A "cold zone" appears in the last 40 m.

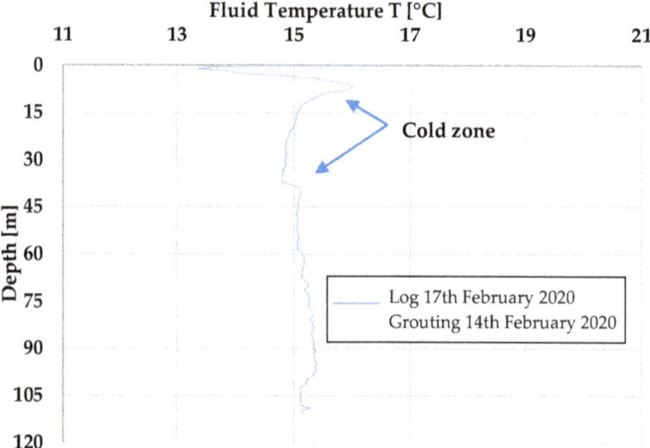

Figure 9. Vertical ground temperature profiles at the TAS-2 BHE during the grout hydration period. A "cold zone" appears in between 13 and 35 m.

In order to perform a reliable distributed TRT, it is necessary to record the temperature values when the hydration reaction is completed and the extra heat effects have vanished. To establish when the TRT experiment can be started, several temperature logs with the submersible sensor have been performed. This procedure is the one described in Figure 10 for the ASC site.

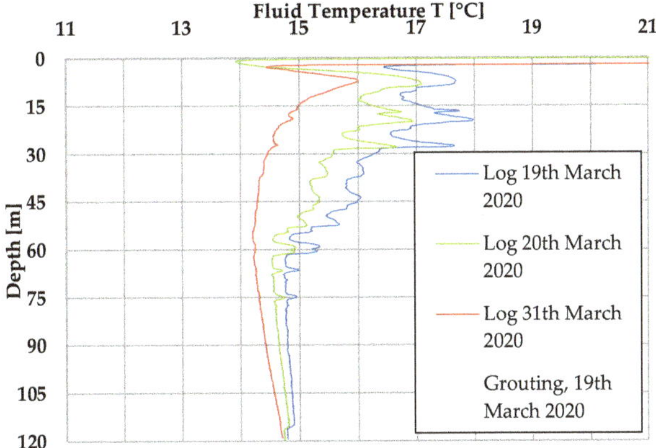

Figure 10. Vertical ground temperature profiles at the ASC-1 pilot BHE during the grout hydration period (19 September profile) as compared to the one at the end of the chemical reaction.

Figure 10 shows that the hydration heat is active in the first days from grouting and that the undisturbed condition is reached in some 12 days (red curve).

Figure 11 shows the measurements in the ASC-1 experiment after the complete decay of any hydration effect. In this case, the distributed TRT experiment lasted 50 h, during which the travelling sensor was inserted 4 times inside the pipe for recording the temperature and pressure. The measurements have been recorded approximately every 0.1 m, and the complete record hence contains thousands of data points. By performing the analysis described by Equations (5)–(7), it was possible to infer the ground thermal conductivity and borehole thermal resistance at each depth position from the slope and intercept of the temperature profile when represented as a function of the logarithm of time. Table 1 shows some 8 of 1200 rows of data recorded during the experiment described in Figure 11. Columns 2 to 6 report the temperatures at several sampling times.

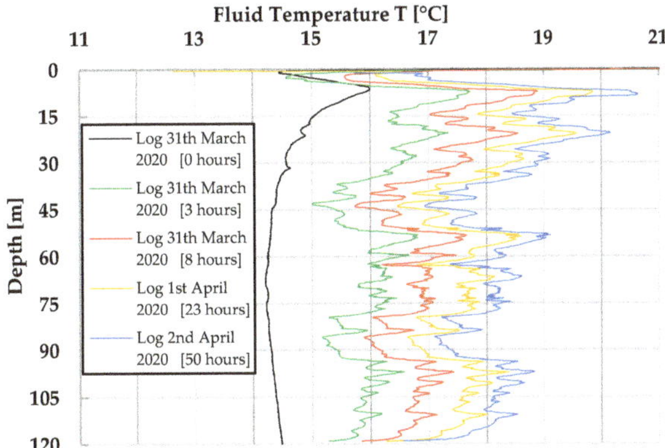

Figure 11. Distributed Thermal Response Test (TRT) experiment at the ASC site after the complete decay of any hydration effect. Vertical temperature profiles along the BHE depth during a 50 h experiment at a constant heat transfer rate.

Table 1. A selection of the dataset containing all the measured temperatures in time and space. Column A and B are the Infinite Line Source (ILS) model estimations for the slope and intercept, respectively (Equation 5), as obtained by log-linear regression for each row. Columns k_{gr} and R_{bhe} are the estimates for the ground and BHE properties as solutions of Equation (7) applied to each row.

Depth	T at 0 h	T at 3 h	T at 8 h	T at 23 h	T at 50 h	A	B	k_{gr}	R_{bhe}
1.4	14.54	14.85	15.94	16.21	16.91	0.682	8.676	2.335	0.150
1.5	14.54	14.85	15.94	16.21	16.91	0.722	8.194	2.206	0.169
1.6	14.54	14.85	15.94	16.21	16.91	0.725	8.098	2.196	0.175
1.7	14.54	14.85	15.94	16.21	16.91	0.726	8.049	2.193	0.178
[—]	[—]	[—]	[—]	[—]	[—]	[—]	[—]	[—]	[—]
118.7	14.47	15.45	16.17	16.81	17.06	0.587	10.089	2.713	0.096
118.8	14.47	15.41	16.12	16.75	16.96	0.567	10.234	2.806	0.093
118.9	14.48	15.33	16.06	16.59	16.85	0.547	10.345	2.909	0.092
119.0	14.47	15.28	15.93	16.51	16.56	0.480	10.926	3.316	0.076
Average	16.01	16.91	17.70	18.23	16.08	0.772	8.965	2.064	0.119

Columns A and B represent the slope m and intercept a of the ILS fitting line (Equation (6)), respectively. Furthermore, columns k_{gr} and R_{bhe} provide estimates of the corresponding quantities at the given depth. Finally, the last row of Table 1 shows the averages of each column.

Figure 12 shows the results of such an analysis. In particular, Figure 12 shows that at a given depth layer (about 65 m), the estimated parameters are significantly different from those at other locations, thus allowing one to guess that that ground layer is characterized by groundwater circulation, which is able to enhance the local heat transfer in terms of a higher (apparent) ground conductivity.

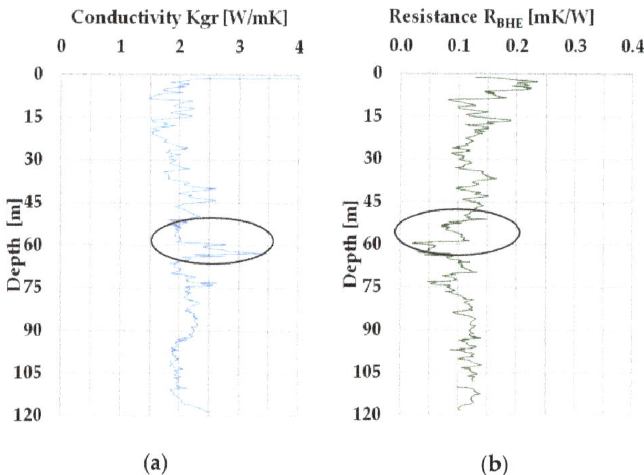

Figure 12. (a) Estimated conductivity k_{gr} and (b) thermal resistance R_{bhe} along the BHE depth as inferred from the ASC-1 measurements. The presence of peak values at the BHE mid depth (circles in figures) can be associated to the presence of groundwater circulation.

Figure 13 finally shows the depth-averaged values of temperatures as a function of the logarithm of time. It can be observed that the experimental points well fit the regression line that is employed for the depth-averaged conductivity estimation. In particular, the fitting line slope (0.772) can be compared to the average of slopes (column A in Table 1, bottom row) as obtained layer by layer.

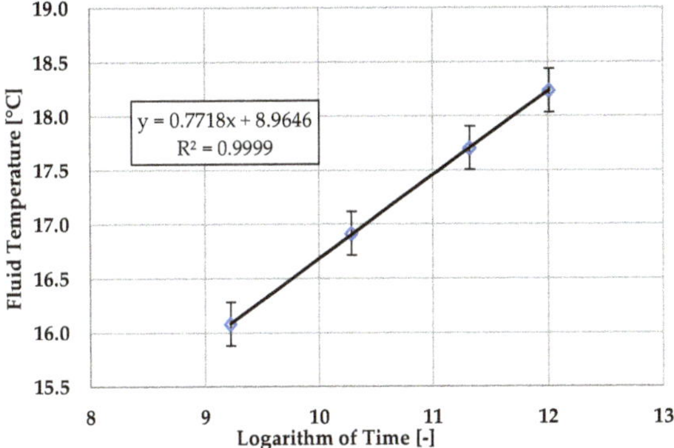

Figure 13. Depth-averaged temperatures as the function of the logarithm of time. The fitting line is employed for a depth-averaged conductivity estimation.

The same analysis can be made in terms of the fitting line intercept (8.965 in Figure 13) and the corresponding average value in Table 1 (column B, bottom row). It is apparent from such a comparison that average k_{gr} and R_{bhe} values as estimated with both methods are very similar, and their agreement is within 1%, thus providing a cross validation of the whole experimental procedure.

6. Conclusions

A series of experiments have been performed in a pilot BHE located in northern Italy during the period 2018–2020. Such measurements have been realized by means of submergible sensors able to record the local temperature along the BHE depth with a spatial step of less than 0.5 m and typically equal to 0.1 m. The temperature profiles along the vertical directions have been employed for a double analysis. The first part of the present investigation was devoted to assessing the effects of the hydration heat released by the grouting cement. It is here demonstrated that the grout chemical reaction can increase the local ground temperature close to the BHE pipes by up to 1.5 °C, and that this effect vanishes after a period ranging from 10 days to 2 weeks.

The same measurements have been employed for detecting the presence of "cold zones" along the borehole heat exchanger, where the local temperature does not change in time during the hydration period, when chemical heat release is expected.

The presence of cold zones is here ascribed to the lack of chemical reactions due to poor grouting or, in other words, to the presence of air, water, or gravel pockets in certain stretches of the BHE volume. In this sense, the present distributed measurement technique offers interesting opportunities for checking the quality of the grouting process in order to assess and forecast the BHE future performance.

The distributed temperature measurements have been finally employed for electrical heating TRT experiments. In this case, pipes are filled by water and one of them if fitted with the electric cable, while the other is used to sink the travelling sensor. It is here confirmed that the technique can be applied with the standard set of equations described by the ILS theory in order to estimate the local (depth related) values of both the ground thermal conductivity and the BHE thermal resistance.

Author Contributions: G.C. and C.P.: measurements; F.M.: hydration parts and results presentation; M.F.: theoretical analysis and general manuscript organization. All authors have read and agreed to the published version of the manuscript.

Funding: Marco Fossa acknowledges the funding by Prin 2017 grant "Heat Transfer and Thermal Energy Storage Enhancement by Foams and Nanoparticles".

Acknowledgments: A.P. and S.M., at Dime, are acknowledged for their contribution in reviewing the present paper and their useful suggestions for improving it.

Conflicts of Interest: The authors declare no conflict of interest.

Nomenclature

Symbol	Variable	Unit
BHE	Borehole Heat Exchanger	-
TRT	Thermal Response Test	-
k_{gr}	the effective ground thermal conductivity	W/(m K)
R_{bhe}	effective borehole thermal resistance	(m K)/W
Q'	heat transfer rate per unit length	W/m
r	radial distance from the line source	M
r_b	borehole radius	M
Fo	Fourier number	-
τ	time	S
T	Temperature	°C
$T_{f,ave}$	average fluid temperature	°C
$T_{gr,\infty}$	undisturbed ground temperature	°C

References

1. Aranzabal, N.; Martos, J.; Steger, H.; Blum, P.; Soret, J. Temperature measurements along a vertical borehole heat exchanger: A method comparison. *Renew. Energy* **2019**, *143*, 1247–1258. [CrossRef]
2. Zeng, H.; Diao, N.; Fang, Z. Efficiency of vertical geothermal heat exchangers in the ground source heat pump system. *J. Therm. Sci.* **2003**, *12*, 77–81. [CrossRef]
3. Luo, J.; Rohn, J.; Bayer, M.; Priess, A. Thermal performance and economic evaluation of double U-tube borehole heat exchanger with three different borehole diameters. *Energy Build.* **2013**, *67*, 217–224. [CrossRef]
4. Sutton, M.G.; Nutter, D.W.; Couvillion, R.J. A ground resistance for vertical bore heat exchangers with groundwater flow. *J. Energy Resour. Technol.* **2003**, *125*, 183–189. [CrossRef]
5. Wagner, V.; Blum, P.; Kübert, M.; Bayer, P. Analytical approach to groundwater influenced thermal response tests of grouted borehole heat exchangers. *Geothermics* **2013**, *46*, 22–31. [CrossRef]
6. IEA ECES, Annex 21. Thermal Response Test. Available online: http://projects.gtk.fi/Annex21/homepage.htm (accessed on 18 May 2020).
7. UNI—Ente Nazionale di Unificazione, UNI 11466:2012. *Sistemi Geotermici a Pompa di Calore—Requisiti per il Dimensionamento e la Progettazione*; Ente Nazionale di Unificazione: Milan, Italy, 2012.
8. Stauffer, F.; Bayer, P.; Blum, P.; Molina-Giraldo, N.; Kinzelbach, W. *Thermal Use of Shallow Groundwater*; CRC Press: Boca Raton, FL, USA, 2013. [CrossRef]
9. Bandos, T.V.; Montero, Á.; Fernández, E.; Santander, J.L.G.; Isidro, J.M.; Pérez, J.; de Córdoba, P.J.F.; Urchueguía, J.F. Finite line-source model for borehole heat exchangers: Effect of vertical temperature variations. *Geothermics* **2009**, *38*, 263–270. [CrossRef]
10. Martos, J.; Montero, A.; Torres, J.; Soret, J.; Martínez, G.; García-Olcina, R. Novel wireless sensor system for dynamic characterization of borehole heat exchangers. *Sensors* **2011**, *11*, 7082–7094. [CrossRef]
11. GEOsniff®Geothermal Monitoring Innovative and Miniaturized. Available online: https://www.enoware.de/en/products/geosniff/ (accessed on 16 May 2020).
12. Aranzabal, N.; Martos, J.; Stokuca, M.; Mazzotti Pallard, W.; Acuña, J.; Soret, J.; Blum, P. Novel instruments and methods to estimate depth-specific thermal properties in borehole heat exchangers. *Geothermics* **2020**, *86*, 101813. [CrossRef]
13. Hellström, G. Ground Heat Storage. Thermal Analysis of Duct Storage Systems: Theory. Ph.D. Thesis, University of Lund, Lund, Sweden, 1991.
14. Paul, N.D. The Effect of Grout Thermal Conductivity on Vertical Geothermal Heat Exchanger Design and Performance. Master's Thesis, South Dakota State University, Brookings, South Dakota, 1996.
15. Zeng, H.; Diao, N.; Fang, Z. Heat transfer analysis of boreholes in vertical ground heat exchangers. *Int. J. Heat Mass Transf.* **2003**, *46*, 4467–4481. [CrossRef]
16. IGHSPA. *Grouting for Vertical Geothermal Heat Pump Systems: Engineering Design and Field Procedures Manual*; Electric Power Research Institute (EPRI)—IGHSPA: Stillwater, Oklahoma, 2000.
17. UNI—Ente Nazionale di Unificazione, UNI 11467:2012. *Sistemi Geotermici a Pompa di Calore—Requisiti per L'installazione*; Ente Nazionale di Unificazione: Milan, Italy, 2012.

18. Verein Deutscher Ingenieure-VDI, VDI 4640 Blatt 2:2019. *Thermal Use of Underground—Ground Source Heat Pump Systems*; Verein Deutscher Ingenieure: Düsseldorf, Germany, 2019.

19. Claesson, J.; Hellström, G. Multipole method to calculate borehole thermal resistances in a borehole heat exchanger. *HVAC R Res.* **2011**, *17*, 895–911. [CrossRef]

20. Lamarche, L.; Kajl, S.; Beauchamp, B. A review of methods to evaluate borehole thermal resistances in geothermal heat-pump systems. *Geothermics* **2010**, *39*, 187–200. [CrossRef]

21. Rees, S. *Advances in Ground-Source Heat Pump Systems*, 1st ed.; Woodhead Publishing: Sawston, UK, 2016; ISBN 9780081003114.

22. Berndt, M.; Kavanaugh, S. Thermal Conductivity of Cementitious Grouts and Impact on Heat Exchanger Length Design for Ground Source Heat Pumps. *HVAC R Res.* **1999**, *5*, 85–96. [CrossRef]

23. Hossein, J.; Seyed, M.A.; Rosen, M.; Pourfallah, M. A comprehensive review of backfill materials and their effects on ground heat exchanger performance. *Sustainability* **2018**, *10*, 4486. [CrossRef]

24. Delaleux, F.; Py, X.; Olives, R.; Dominguez, A. Enhancement of geothermal borehole heat exchangers performances by improvement of bentonite grouts conductivity. *Appl. Therm. Eng.* **2012**, *33*, 92–99. [CrossRef]

25. Alrtimi, A.A.; Rouainia, M.; Manning, D.A.C. Thermal enhancement of PFA-based grout for geothermal heat exchangers. *Appl. Therm. Eng.* **2013**, *54*, 559–564. [CrossRef]

26. Chen, F.; Mao, J.; Chen, S.; Li, C.; Hou, P.; Liao, L. Efficiency analysis of utilizing phase change materials as grout for a vertical U-tube heat exchanger coupled ground source heat pump system. *Appl. Thermal Eng.* **2018**, *130*, 698–709. [CrossRef]

27. Philippacopoulos, A.J.; Berndt, M.L. Influence of debonding in ground heat exchangers used with geothermal heat pumps. *Geothermics* **2001**, *30*, 527–545. [CrossRef]

28. Chen, S.; Mao, J. Quantitative evaluation of improper backfilling in vertical borehole. *J. Therm. Sci. Technol.* **2016**, *11*, JTST00018. [CrossRef]

29. Suibert Oskar Seibertz, K.; Händel, F.; Dietrich, P.; Vienken, T. On the use of hydration heat for quality management of borehole heat exchanger grouting. In Proceedings of the EGU General Assembly Conference Abstracts, EGU General Assembly, Vienna, Austria, 17–22 April 2016; p. EPSC2016-12711. Available online: https://ui.adsabs.harvard.edu/abs/2016EGUGA.1812711S (accessed on 18 May 2020).

30. Mogensen, P. Fluid to duct wall heat transfer in duct system heat storages. *Doc. Swed. Counc. Build. Res.* **1983**, *16*, 652–657.

31. Gehlin, S. Thermal Response Test Method Development and Evaluation. Ph.D. Thesis, Lulea University of Technology, Lulea, Sweden, 1996.

32. Austin, W.A. Development of an In Situ System for Measuring ground Thermal Properties. Ph.D. Thesis, Oklahoma State University, Stillwater, Oklahoma, 1998.

33. Acuna, J.; Mogensen, P.; Palm, B. Distributed thermal response test on a U-pipe borehole heat exchanger. In Proceedings of the 11th International Conference on Energy Storage, EFFSTOCK, Stockholm, Sweden, 16 June 2009.

34. Fossa, M.; Rolando, D.; Pasquier, P. Pulsated Thermal Response Test experiments and modelling for ground thermal property estimation. In Proceedings of the International Ground Source Heat Pump Association Research Conference, Stockholm, Sweden, 18–20 September 2018; pp. 220–228. [CrossRef]

35. Morchio, S.; Fossa, M. On the ground thermal conductivity estimation with coaxial borehole heat exchangers according to different undisturbed ground temperature profiles. *Appl. Therm. Eng.* **2020**. [CrossRef]

36. Fossa, M.; Priarone, A.; Silenzi, F. Superposition of the Single Point Source Solution to Generate Temperature Response Factors for Geothermal Piles. *Renew. Energy* **2020**, *145*, 805–813. [CrossRef]

37. Franco, A.; Conti, P. Clearing a Path for Ground Heat Exchange Systems: A Review on Thermal Response Test (TRT) Methods and a Geotechnical Routine Test for Estimating Soil Thermal Properties. *Energies* **2020**, *13*, 2965. [CrossRef]

38. Ingersoll, L.R.; Zobel, O.J.; Ingersoll, A.C. *Heat Conduction with Engineering, Geological, and Other Applications*; McGraw-Hill: New York, NY, USA, 1954.

39. Fossa, M. Correct Design of Vertical BHE Systems through the Improvement of the Ashrae Method. *Sci. Technol. Built Environ.* **2016**, *23*, 1080–1089. [CrossRef]

40. Eskilson, P. Thermal Analysis of Heat Extraction Boreholes. Ph.D. Thesis, Lund University of Technology, Lund, Sweden, 1987.

41. Hu, J.; Ge, Z.; Wang, K. Influence of cement fineness and water-to-cement ratio on mortar early-age heat of hydration and set times. *Constr. Build. Mater.* **2014**, *50*, 657–663. [CrossRef]
42. Kavanaugh, S.P. Field Tests for Ground Thermal Properties—Methods and Impact on Ground-Source Heat Pump Design. *ASHRAE Trans.* **2000**, *106*, 851–855.
43. ASHRAE. *ASHRAE Fundamentals Handbook*; ASHRAE: Atlanta, GA, USA, 2007; Chapter A32.
44. Geoportale Regione Emilia-Romagna. Available online: https://geoportale.regione.emilia-romagna.it/ (accessed on 22 May 2020).

Article

Evaluation of Heat Transfer Performance of a Multi-Disc Sorption Bed Dedicated for Adsorption Cooling Technology

Marcin Sosnowski

Faculty of Science and Technology, Jan Dlugosz University in Czestochowa, Czestochowa, Armii Krajowej 13/15 42-200, Poland; m.sosnowski@ujd.edu.pl; Tel.: +48-34-36-12179

Received: 10 November 2019; Accepted: 6 December 2019; Published: 8 December 2019

Abstract: The possibility of implementing the innovative multi-disc sorption bed combined with the heat exchanger into the adsorption cooling technology is investigated experimentally and numerically in the paper. The developed in-house sorption model incorporated into the commercial computational fluid dynamics (CFD) code was applied within the analysis. The research allowed to define the design parameters of the proposed type of the sorption bed and correlate them with basic factors influencing the performance of the sorption bed and its dimensions. The designed multi-disc sorption bed is characterized by great scalability and allows to significantly expand the potential installation sites of the adsorption chillers.

Keywords: adsorption; cooling technology; chiller; heat exchanger; waste heat utilization; CFD

---

1. Adsorption Cooling Technology

1.1. Environmental Demands

Energy and environment-related issues are strongly inter-related and are becoming the most important and popular topics in research activities nowadays [1]. The European Union's concerns are focused on energetic efficiency as the most effective method of reducing primary energy consumption. It directly lowers the emissions of harmful substances into the atmosphere and, consequently, improves air quality. An important element of activities aimed at increasing energetic efficiency is the maximum use of heat that has so far been released into the atmosphere and its conversion into usable energy. Moreover, the two main expectations defined in the concept of sustainable development are the reduction of energy consumption from non-renewable sources and the effective utilization of low-grade thermal energy originating from industrial waste heat, solar power, cogeneration or exhaust gases from internal combustion engines [2].

On the other hand, the demand for cooling in the industrial and residential sectors is increasing. The mechanical vapor-compression air-conditioning systems are commonly used to meet such a demand for cooling and their popularity results from high coefficients of performance (COP), small sizes, and low weights. Unfortunately, they vastly contribute to the global warming and ozone layer depletion because of the usage of refrigerants such as chlorofluorocarbon, hydro-chlorofluorocarbon, or hydrofluorocarbons [3]. Therefore, the development and wide-spread usage of an alternative solution to the conventional cooling systems is a necessity.

1.2. Adsorption Chillers

The devices, which are capable to generate chill on the ground of adsorption cooling technology can be the perfect solution to the aforementioned issues [4–6]. The adsorption chillers can be successfully powered with low-grade heat, which can be obtained from renewable sources, e.g., solar radiation [6], geothermal energy [7], or industrial waste heat [8–11]. They can be applied in combined cooling, heating, and power (CCHP) systems in numerous industrial and commercial applications [12] as well as in sustainable building air-conditioning using solar energy as heat source [1]. The production of chill, practically without the use of electricity, by absorption chillers, has a particular significance in summer, during the peak period of demand for electricity from the network to supply conventional vapor-compression refrigeration systems. In addition, the adsorbents in the adsorption cooling technology are porous materials like silica gel, zeolites, and activated carbons, which are environmentally friendly with no impact to the atmosphere [13]. The adsorbates are natural coolants such as water, ethanol, methanol, CO_2, or ammonia. These working pairs in adsorption cooling technology are characterized by zero global warming potential (GPW) and ozone depletion potential (ODP) [14].

Other important benefits of adsorption chillers in comparison to conventional vapor-compression systems are driving these devices with renewable or waste heat source of temperature as low as 50 °C [7], which directly leads to a reduction in CO_2 emissions and pollution [15], almost zero electricity consumption [6], no moving parts resulting in high reliability [16], simple control and maintenance [3]. Moreover, on the markets of the Middle and the Far East, adsorption technology is intensively developed because of the possibility of desalination of seawater along with the production of cooling energy [17]. But the widespread application of adsorption chillers is limited by the following shortcomings of adsorption cooling technology: low coefficient of performance [18], large weight and volume [16], intermittent cooling [14], high initial procurement cost [14], and exploitation under vacuum conditions [2].

1.3. Design and Operation

The adsorption phenomenon is the binding of the adsorbate to the surface of the adsorbent. The adsorption cooling technology utilizes the physical sorption phenomenon, where the molecules of adsorbate are bound to the surface of the porous adsorbent by Van-der-Waals forces [16]. These are intermolecular interactions between two dipoles, one of which is induced by another permanent dipole [15]. The combination of subsequent adsorption and desorption cycles is used to produce the cooling effect in the adsorption cooling technology.

The single-bed, single-stage adsorption chiller consists of an evaporator, condenser, valves separating adjacent devices, and at least one sorption bed [19] as shown in Figure 1a. The advances in the design of evaporators and condensers are described in [20,21]. The sorption bed is designed as a hermetic container with a built-in heat exchanger, which transfers heat between the sorbent and the heating/cooling medium (usually water). More than one sorption bed can be applied in order to reduce the intermittency of chill production and assure heat and mass recuperation. The granular packed adsorbent bed design is usually used in the adsorption cooling systems [12] because it is characterized by high mass transfer performance due to the high permeability level and developed specific surface of the adsorbents.

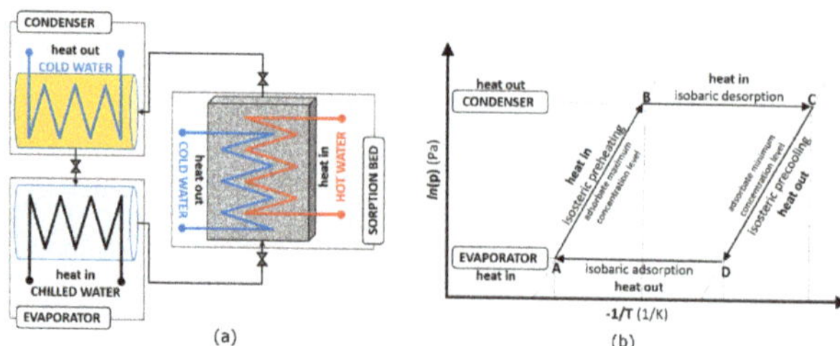

Figure 1. Basic, single-bed, single-stage adsorption cooling system: (**a**) chiller design; (**b**) Clapeyron diagram.

The ideal adsorption refrigeration cycle is typically expressed by the Clapeyron diagram [22] with respect to the isosteres of adsorbent–adsorbate pair as shown in Figure 1b. The sorption bed operates between the condenser pressure and the evaporator pressure as well as the minimum and maximum adsorbate concentration levels. The Clapeyron diagram (Figure 1b) illustrates the four ideal thermodynamic steps occurring in the bed i.e., isosteric preheating (A–B), isobaric desorption (B–C), isosteric precooling (C–D), and isobaric adsorption (D–A) [23].

The absorptivity of the adsorbent is directly proportional to the pressure. The adsorption conducted under high pressure would be optimal because of the use of the absorptive capacity of the bed; however, the adsorbate vapor temperature is strictly correlated with the pressure. In the evaporator, intensively evaporating adsorbate under reduced pressure rejects heat from the chilled water, thus generating the cooling power. That is why the chilled water temperature is related to the evaporator pressure; the lowest chilled water temperatures are achievable at low adsorption pressures. It is, therefore, necessary to balance between a low temperature of the obtained chill and the efficiency of the device.

1.4. Literature Review

The literature reports valuable examples of research concerning the adsorption cooling technology. Sakoda and Suzuki performed a fundamental study on solar-powered, silica gel–water adsorption cooling system [24]. They also analyzed the transport of heat and adsorbate in a small-scale unit on the ground of both experimental and numerical research [25]. Numerous research report the potential for applying different heat transfer enhancements techniques to improve adsorption bed thermal performance. One of them is based on replacing the gaseous voids of low thermal conductivity between the individual grains of sorbent with the glue of better thermal properties as was investigated in [26–31]. The obtained results confirmed the beneficial influence of this technique on the COP. Aristov et al. [32] optimized the adsorption dynamics with the application of loose-grain sorbent. The insertion of metallic additives to the sorbent was investigated in [12,33]. Another approach based on a polydispersed composition of the sorption bed was analyzed by Girnik and Aristov [34] and Demir et al. [35]. Alam et al. [36] numerically investigated the heat exchanger design effect on the performance of closed cycle, two-bed adsorption cooling systems. The authors defined a non-dimensional switching frequency and studied the effect of heat exchanger design parameters on the system performance. The modifications of the heat exchanger design were also investigated in [18,37,38]. Ilis et al. [39] performed very interesting research concerning the innovative star fin type adsorbent bed design. Two dimensional numerical analysis allowed to determine the optimum geometrical parameters and find the best specific cooling performance value. Moreover, the influence of metal additives on the performance of the adsorption chiller was investigated. The finned tube heat exchanger was analyzed in [40] and flat-tube heat exchanger was examined in [41]. A 2D coupled heat and mass transfer model was used in [42] to analyze the performance of both finless and finned tube-type adsorbent beds during only the desorption mode. A significant enhancement in the heat transfer was obtained using a finned tube adsorbent bed. The effect of fin design parameters on the heat transfer inside the bed was also investigated using four different fin configurations.

Researches evaluated different adsorption pairs [43,44] but silica gel/water has shown significant advantages in terms of thermal performance and environmental impact [12]. Water vapor as adsorbate is characterized by excellent thermo-physical properties of high latent heat of evaporation, high thermal conductivity, low viscosity, and thermal stability in a wide range of operating temperatures. Silica gel is a synthetically obtained porous form of silicon oxide (SiO_2) characterized by high chemical resistance. The advantages of silica gel as adsorbent are also revealed in high adsorption/desorption rate, low regeneration heat, good long-term stability, and minimal hysteresis [45]. The analysis of the thermal behavior of devices using silica gel shows that their performance is very sensitive to heat and mass transfer rates inside the adsorbent beds [46]. According to [47], the lowest temperature needed to regenerate the adsorption chiller bed is required for the silica gel–water pair. Therefore, this sorption pair opens up a number of chiller's possible applications because of the low temperature of the required heat source.

Apart from experimental research, several numerical methods were applied in studies on adsorption cooling technology. Krzywanski et al. [48–50] successfully utilized the genetic algorithms, neural networks, and AI approach for adsorption chiller work cycle analysis. The rapid development of high-performance computing (HPC) resulted in the increasing scope of computational fluid dynamics (CFD) application, also in the area of adsorption technology. Papakokkinos et al. [51] presented a generalized three-dimensional CFD model based on the unstructured meshes. Dynamic conjugate simulations of the packed bed and the heat exchanger allowed to study the influence of sorption bed geometry on the reactor performance. The effect of the adsorbed mass spatial distribution on the desorption phase was also discussed and the strong impact of the solid volume fraction, fin length, and fin thickness on the heat transfer was demonstrated. A similar parametric study concerning finned tube heat exchanger was performed in [37]. Detailed analysis of flow characteristics and heat transfer within a packed bed of sorbent using CFD technique were also performed in [52–55]. This research tool was also applied in studies dealing in adsorption dynamics of cylindrical silica gel particles [56], where authors adopted a three-dimensional finite volume method for solving the coupled energy and mass diffusion equations.

1.5. The Main Aim of the Work

The key challenges in the development of the adsorption cooling technology are the low performance and large dimensions of the adsorption chillers. Enhancing the adsorption kinetics by improving heat and mass transfer is necessary in order to improve the adsorption chiller performance [12]. Both, performance and dimensions of the device are dependent on the design of a heat exchanger constituting the adsorption bed [57,58], which is the most important part of the chiller [14]. That is why the advances in the heat exchanger design will directly contribute to diminishing the drawbacks of these environmentally friendly devices.

Therefore, the aim of this research is to investigate the possibility of implementing the innovative multi-disc sorption bed in an adsorption cooling technology in order to increase the COP of adsorption cooling devices by intensifying the heat transfer in a sorption bed as well as open up the possibilities of wider use of adsorption chillers because of the scalable and flat design of the bed.

2. Research Object

2.1. Multi-Disc Sorption Bed Design

The innovative construction of a multi-disc sorption bed proposed by the author and depicted in Figure 2 is investigated in this research. In contrast to the commonly-applied designs, in a multi-disc bed, the sorbent is placed in many separate disc-shaped packets, and the cooling/heating water washes the packets of sorbet from the outside transferring heat. The adsorbate vapor flows through the fixing net into the sorbent packets penetrating them. The fixing net holds the granular sorbent inside the disc-shaped packets.

The proposed construction allows to place the device e.g., in the ceiling of the building or integrate it with solar panels, which will supply the device with the necessary heat. Such a solution allows to significantly expand the potential installation sites of the adsorption chillers and thus reduce one of the main disadvantages of these devices, which is the need to save a large space for the installation of the adsorption chiller. Another advantage of the proposed solution is its potential for scalability consisting of adjusting the number of sorbent discs to the expected cooling capacity of the device or the possibility of installing two or more multi-disc sorption beds with sorbent packages one above another with the space between them being a vapor collector.

The analysis of the commercially available adsorption cooling equipment and the literature review showed that the presented solution is not the current state of the research field. Therefore, a patent application was prepared for the Polish Patent Office on the basis of the presented multi-disc sorption bed design.

Figure 2. Sectional view of the investigated multi-disc sorption bed. (**a**) disc-shaped sorbent packets; (**b**) fixing net; (**c**) heat exchanger main body.

2.2. Inlet/Outlet Manifolds

In the subsequent stages of the implementation works, it was necessary to build a lab-scale prototype of the multi-disc sorption bed. The crucial step in this stage was to design the inlet/outlet manifolds of cooling/heating water in order to incorporate the heat exchanger into the operational test stand. Therefore, three variants of the inlet/outlet manifold geometry depicted in Figure 3 have been investigated with the use of numerical methods.

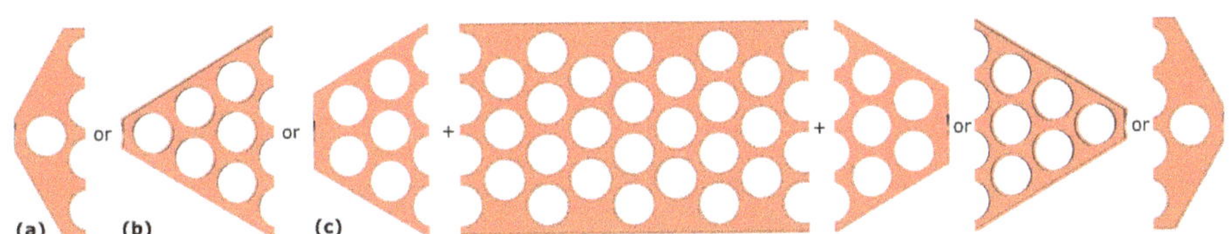

Figure 3. Variants of investigated inlet/outlet manifolds. (**a–c**) represent three variants of the inlet/outlet manifold geometry.

The most uniform spatial distribution of heating medium velocity within the main part of the multi-disc sorption bed and the largest temperature difference between inlet and outlet was obtained with variant (a) depicted in Figure 3 and therefore this inlet/outlet manifold geometry was used in the lab-scale prototype of the device for further analysis.

2.3. Lab-Scale Prototype

The physical prototype of the multi-disc sorption bed depicted in Figure 4 was constructed in the lab-scale. The detailed dimensions of the prototype are shown in Figure 5.

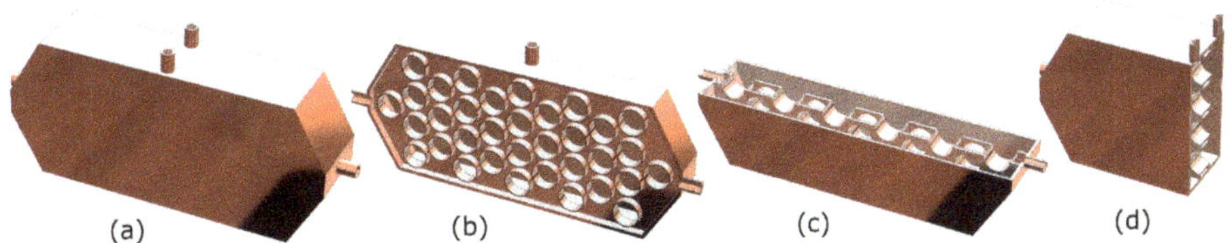

Figure 4. The prototype of the multi-disc sorption bed. (**a**) 3D view; (**b–d**) cross-section views.

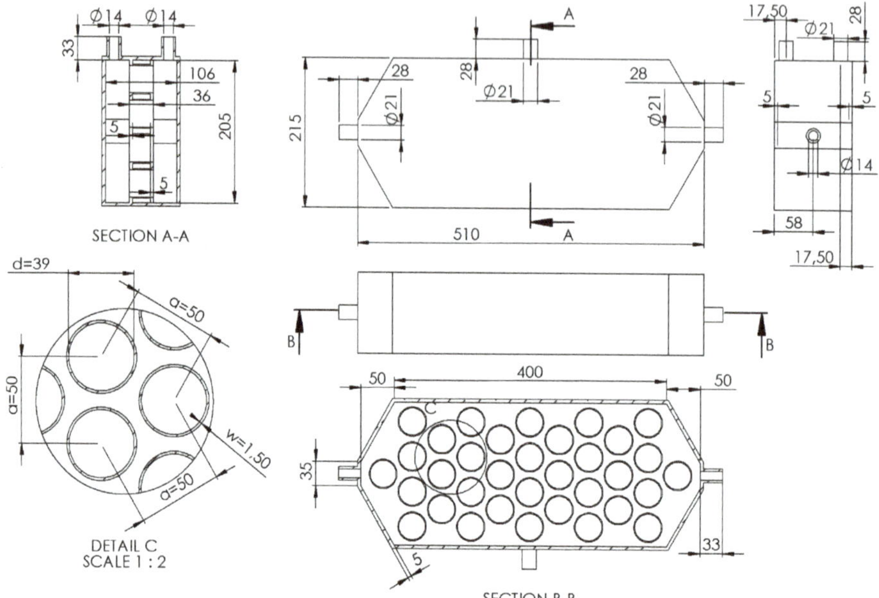

Figure 5. The dimensions (in mm) of the multi-disc sorption bed prototype.

The multi-disc sorption bed prototype is made of copper and the design allows to deliver water liquid, water vapor, or any other fluid medium through the connecting pipes to the two separate and water-tight volumes. The sorbent can be placed in the 34 disc-shaped packets and is secured with the fixing net (not shown in Figures 4 and 5).

3. Research Methods

3.1. Experimental Research

In the first stage of the research, the physical prototype of the multi-disc sorption bed was experimentally tested as a water–water heat exchanger in order to evaluate its heat transfer efficiency not being influenced by the sorption processes. Therefore, it was connected to the hot water supply of adjustable temperature and an open loop of cold water supply as depicted in Figure 6.

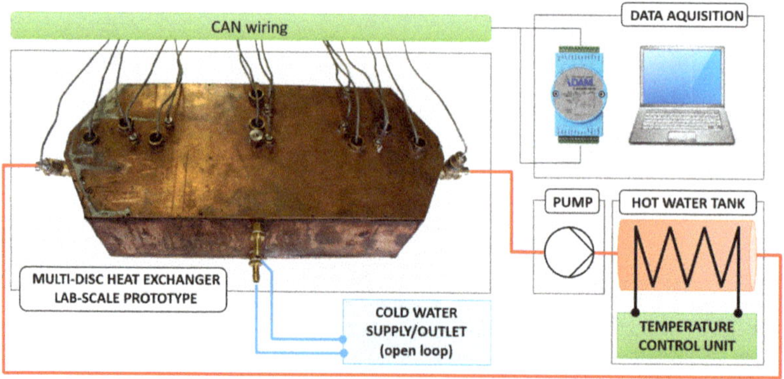

Figure 6. The experimental setup.

The prototype was equipped with PT-100 two-wire temperature sensors, with the measuring range of −50 to +150 K and the measuring accuracy of ±0.1 K. The measuring probes were installed in the selected packets as well as outside the packets in the water jacket of the bed as depicted in Figure 7. The additional probes were installed in the inlets and outlets of both hot and cold water supplies.

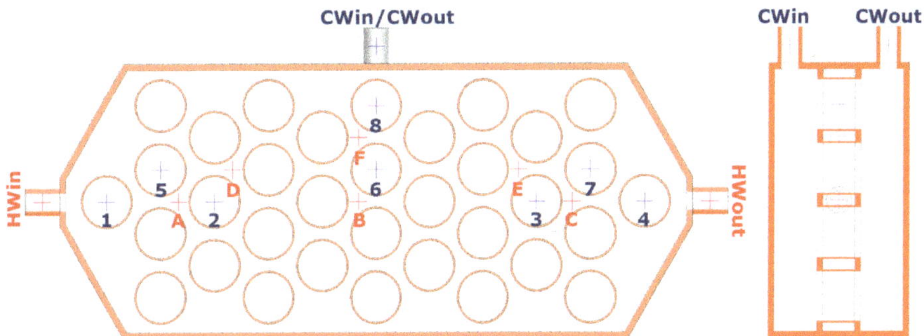

Figure 7. The temperature measuring points: A–F—hot water probes; 1–8—cold water probes.

The ADAM-4015 RTD module manufactured by Advantech was used to acquire the temperature values from individual measurement points with a frequency of 1 Hz. The turbine water flow meters of the measurement range of 0.008–0.2 dm^3/s and the measurement accuracy of ±2% were used in order to determine the mass flow rate of both hot and cold water.

3.2. Numerical Research

3.2.1. CFD Tool

The application of numerical simulation tools calibrated with experimental measurements is a practical and cost-effective approach for energy simulation analysis [59]. CFD is the valuable simulation tool for the design of adsorption cooling systems and many researchers report the potential of using validated simulation models to investigate the performance of adsorption chillers [12]. CFD has been successfully applied in numerous research concerning conjugate heat transfer [53] as well as adsorption cooling and desalination technology [18,37]. Moreover, CFD allows for rapid prototyping and reliable analysis of several design configurations. The significance of CFD and the scope of its application have been increasing because of the rapid development of high-performance computing as well as cloud computing. Therefore, the commercial CFD package, ANSYS Fluent 2019 R1, was applied to carry out the numerical research. Its algorithm constrains the mass conservation of the velocity field by solving a pressure equation derived from the continuity and momentum equations. The nonlinear and coupled governing equations are solved iteratively.

In addition, an in-house code enhancing the solver capabilities of modeling sorption processes and described in Section 3.2.4. was applied within the carried out research.

3.2.2. Computational Domain and Discretization

The associative CAD model corresponding to the physical prototype of the multi-disc sorption bed was developed within the research. The model was parametrized and served as an input for the mesh generator.

In the first stage of the research, the prototype was tested as a water–water heat exchanger in order to validate the CFD model and therefore it consisted of three subdomains representing one solid region (copper multi-disc sorption bed) and two fluid regions (hot water and cold water). In the second stage, apart from the heat transfer and flow processes, the sorption phenomenon in the multi-disc bed was modeled using the developed in-house code. For that reason, the computational domain was modified in order to take into account the sorbent and consisted of two solid regions (copper multi-disc sorption bed, disc-shaped sorbent packets) and two fluid regions (water, water vapor) as depicted in Figure 8.

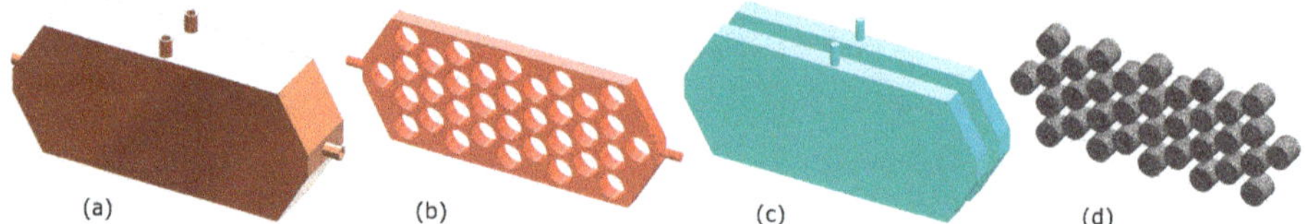

Figure 8. The computational domain consisting of four subdomains. (**a**) Multi-disc sorption bed; (**b**) water; (**c**) water vapor; (**d**) sorbent.

The accuracy and solution time are the two critical issues in CFD and both are highly dependent on the computational domain discretization. Different types of mesh elements are needed to deliver optimal performance in resolving different geometries and flow regimes [60]. Therefore, the MOSAIC® meshing was implemented in order to maintain the layered elements on the boundary layers and fill the rest of the volume with high-quality polyhedral elements (Figure 9). Polyhedral cells consume less memory and computing time in comparison to tetrahedral elements. Moreover, they also have many neighbors, so gradients can be better approximated and layered polyhedral prisms can be applied on the boundaries to efficiently capture the boundary layer on no-slip walls. The mesh is fully conformal, which ensures that the nodes on both sides of the interface between fluid and solid regions match to each other. Such an approach assures no interpolation at the interface, which contributes to the reduction of computational time and ensures higher accuracy of the solution.

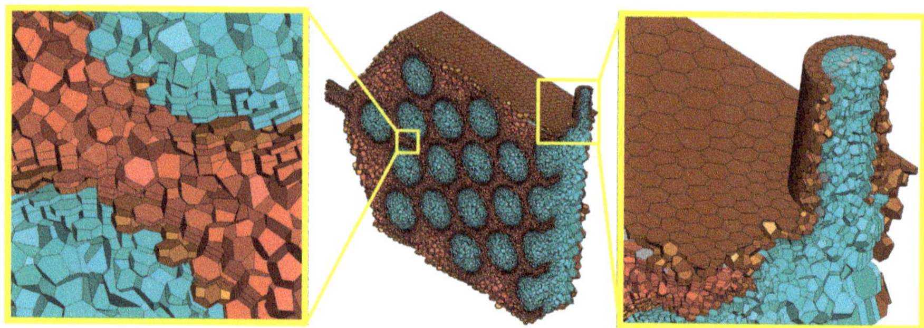

Figure 9. The computational domain discretization.

Mesh dependency studies based on the grid convergence index (*GCI*) were carried out in order to estimate the numerical accuracy resulting from the mesh resolution. The *GCI* is recommended by the Fluids Engineering Division of the American Society of Mechanical Engineers (ASME) to estimate the discretization error and was successfully applied in many research [61–64]. Three numerical meshes of different resolutions were generated for the analyzed geometry in order to estimate the discretization error with *GCI*.

The mean relative cell size was defined as the ratio of the average cell size h, defined in the Equation (1), and the characteristic dimension of the multi-disc sorption bed a, which is the distance between the centers of the two adjacent disc-shaped packets (Figure 5).

$$h = \sqrt[3]{\frac{1}{N} \sum_{i=1}^{N} (\Delta V_i)},$$

(1)

where:

ΔV_i—volume of the ith cell;

N—total number of cells in the computational domain;

Initially, the mesh consisting of elements of mean relative cell size equal to 8.44×10^{-2} was generated and then meshes of mean relative cell sizes decreased to 6.46×10^{-2} and 4.72×10^{-2} were prepared by scaling the initial mesh.

The *GCI* was calculated in order to minimize the discretization error according to the procedure described in [65]. The grid refinement factor r (2) was calculated based on the representative mesh size h (1) as:

$$r = \frac{h_{coarse}}{h_{fine}}, \tag{2}$$

Moreover, the following assumption was made:

$$h_1 < h_2 < h_3; \; r_{21} = \frac{h_2}{h_1}; \; r_{32} = \frac{h_3}{h_2}, \tag{3}$$

The order of convergence p was calculated based on Equation (4) [65]:

$$p = \frac{\left| ln\left|\frac{\varepsilon_{32}}{\varepsilon_{21}}\right| + ln\left(\frac{r_{21}^p - 1 \cdot sgn\left(\frac{\varepsilon_{32}}{\varepsilon_{21}}\right)}{r_{32}^p - 1 \cdot sgn\left(\frac{\varepsilon_{32}}{\varepsilon_{21}}\right)}\right)\right|}{ln(r_{21})}, \tag{4}$$

where:

$$\varepsilon_{32} = \phi_3 - \phi_2; \; \varepsilon_{21} = \phi_2 - \phi_1, \tag{5}$$

ϕ_k denotes the value of the variable important to the objective of the simulation study for the solution obtained with the kth mesh. The logarithmic mean temperature difference (*LMTD*) was selected as the above-mentioned variable.

The approximate relative error was calculated on the basis of Equation (6).

$$e_a^{21} = \left|\frac{\phi_1 - \phi_2}{\phi_1}\right| \tag{6}$$

Finally, the *GCI* was determined using Equation (7) provided in [65]:

$$GCI_{21} = \frac{1.25 \cdot e_a^{21}}{r_{21}^p - 1} \tag{7}$$

All the values of the above-defined quantities are listed in Table 1. The obtained results indicate the mesh convergence with the *GCI* equal to 1.03%.

Table 1. Mesh parameters and values of quantities calculated based on Equations (1)–(7).

h/a (-)	N (-)	*LMTD* (K)	H (-)	R (-)	ε (-)	$\varepsilon_{i+1}/\varepsilon_I$ (-)	P (-)	e_a (%)	*GCI* (%)
$4.7 \cdot 10^{-2}$	888 694	25.84	2.3603	1.3680	−0.31				
$6.5 \cdot 10^{-2}$	347 133	25.53	3.2289	1.3067	−0.10	converged	2.867	1.20%	1.03%
$8.4 \cdot 10^{-2}$	155 590	25.43	4.2192	-	-				

3.2.3. Boundary Conditions and Model Settings

The first stage of numerical investigations aimed to validate the multi-disc sorption bed model operating as a water-water heat exchanger with the obtained experimental data. The heat transfer efficiency of the prototype was tested under three different R_{MFR} ratios defined with Equation (8) and equal to 1.00, 1.33, and 1.66. The cold water inlet temperature was 294.1 K and hot water inlet temperature was 336.2 K. The above data was defined based on the experimental research.

$$R_{MFR} = \frac{\dot{m}_{HW}}{\dot{m}_{CW}}, \tag{8}$$

where:

$\dot{m}_{HW}$—hot water mass flow rate (kg/s);
$\dot{m}_{CW}$—cold water mass flow rate (kg/s);

In the second stage of the research, apart from the heat transfer and flow processes, the desorption process was investigated as the analysis of temperatures inside the bed during the desorption is important to show the spatial distributions of the main parameters of the chiller [2]. The boundary conditions corresponded with the final stage of the desorption phase of the adsorption chiller working cycle, therefore, the water of 343 K was the heating fluid and the mass-flow-inlet boundary condition of mass flow rate equal to 0.005 kg/s was assigned. The pressure-outlet boundary condition was defined on the outlet from the heating water subdomain. The heat transfer between fluid and solid subdomains was calculated as conjugate heat transfer. Parameters of the applied materials are defined in Table 2.

Table 2. Materials parameters.

Material	Density (kg·m^{-3})	Specific Heat (J·kg^{-1}·K^{-1})	Thermal Cond. (W·m^{-1}·K^{-1})	Viscosity (kg·m^{-1}·s^{-1})
water (liquid)	998.2	4182	0.6	1.003·10^{-3}
water (vapor)	0.5542	f(T)	0.0261	1.34·10^{-5}
silica gel	800	924	0.18	-
copper	8978	381	387.6	-

The CFD solver was configured as pressure-based and the analysis was performed for a steady state. The standard k-ε viscous model was applied along with the enhanced wall treatment. Pressure–velocity coupling by the COUPLED algorithm was used as a solution method. Least squares cell-based spatial discretization was chosen in case of gradients, second order in case of pressure, second-order upwind in case of momentum, as well as energy and first-order upwind in case of turbulent dissipation rate. The model convergence was defined on the basis of qualitative and quantitative monitoring of residuals as well as the thermodynamic stability of the model. The developed in-house model described in the next subsection was used in order to model the sorption phenomenon.

3.2.4. Sorption Modeling

The commercial ANSYS Fluent tool used in this research, dedicated to CFD analysis, does not allow for sufficient consideration of the aspects related to heat and mass exchange during sorption processes. Therefore, it was necessary to expand the available models by using the in-house algorithm implemented as a user-defined functions (UDF). The algorithm developed within the research constitutes a novelty in 3D modelling of the sorption processes in adsorption cooling technology. The UDF created within the framework of this research is a program written in the C programming language. In order to effectively incorporate it into the solver code, it was developed using the Microsoft Visual Studio compiler. It was decided to use the compiled UDF because compared to the interpreted UDF, it is characterized by faster operation, no limitation on the programming language, the ability to call functions written in other programs, and the ability to run executable files. Especially the first of the above-mentioned features, i.e., faster operation, was crucial for the choice of the UDF type.

Numerical modeling of sorption processes requires an extensive approach to mathematical description of heat exchange in a sorbent bed, as it is necessary to take into account the fluctuations in the local intensity of heat production or consumption during the exothermic adsorption or endothermic desorption process, respectively. The above-mentioned intensity of heat production or consumption in the bed depends directly on the local temperature of the sorbent, which in turn is closely correlated with the design and operating parameters of the heat exchanger. In order to take these factors into account in the model, it was decided to modify the source term in the energy equation. The source term enables to describe the exo- and endothermic character of sorption processes and, more importantly, their two-way coupling with the thermal-flow field calculated by the solver.

The intensity of sorption processes in the developed model depends on the local temperature of the sorbent by the function allowing to generate the temperature value in a given numerical mesh element by the solver's algorithm. Then the heat source is calculated as a function of the intensity of sorption depending on the local temperature obtained in the first step. Subsequently, the derivative of the heat source with respect to temperature is calculated and the value of the volumetric heat source is assigned to the function returned to the algorithm of the solver.

The mathematical dependence of the sorption intensity and the sorbent temperature was determined as a polynomial function of coefficients defined and validated during previous studies concerning heat transfer in the sorption beds [18,37].

4. Results and Discussion

4.1. Heat Transfer Efficiency

The logarithmic mean temperature difference (*LMTD*) defined in Equation (9) is the average temperature difference between the hot and cold fluids [66] and therefore it was used to determine the efficiency of the multi-disc sorption bed operating as a crossflow water–water heat exchanger.

$$LMTD = \frac{\Delta T_{inlet} - \Delta T_{outlet}}{\ln\left(\frac{\Delta T_{inlet}}{\Delta T_{outlet}}\right)} = \frac{(HW_{in} - CW_{in}) - (HW_{out} - CW_{out})}{\ln\left(\frac{HW_{in}-CW_{in}}{HW_{out}-CW_{out}}\right)}, \tag{9}$$

where:

HW_{in}—hot water inlet temperature (K);
HW_{out}—hot water outlet temperature (K);
CW_{in}—cold water inlet temperature (K);
CW_{out}—cold water outlet temperature (K);

According to [66], the *LMTD* calculated for the crossflow heat exchanger has to be corrected with the factor *F* read from the graph presented in [66] (page 52) and based on the coefficients *P* and *R* defined in Equation (10).

$$P = \frac{CW_{out} - CW_{in}}{HW_{in} - CW_{in}}, \quad R = \frac{HW_{in} - HW_{out}}{CW_{out} - CW_{in}}, \tag{10}$$

The *LMTD*, *P*, *R*, *F*, and corrected *LMTD* for the multi-disc sorption bed operating as a crossflow water–water heat exchanger calculated based on both the experimental and CFD results for the analyzed R_{MFR} are presented in Table 3. The correction factor *F* is close to 1 for all analyzed R_{MFR} ratios, which indicate the very good efficiency of the investigated innovative multi-disc construction.

Table 3. The obtained logarithmic mean temperature difference (*LMTD*), *P*, *R*, *F*, and corrected *LMTD* values.

Analyzed Case	LMTD (K)	P (-)	R (-)	F (-)	F×LMTD (K)
R_{MFR} = 1.00; EXP	27.19	0.306	1.000	0.975	26.51
R_{MFR} = 1.00; CFD	25.84	0.326	1.017	0.970	25.07
R_{MFR} = 1.33; EXP	27.47	0.340	0.776	0.985	27.06
R_{MFR} = 1.33; CFD	26.70	0.361	0.742	0.985	26.30
R_{MFR} = 1.66; EXP	28.23	0.363	0.588	0.985	27.80
R_{MFR} = 1.66; CFD	27.16	0.384	0.598	0.980	26.62

The temperatures obtained during the experimental research were compared with the CFD results in order to validate the numerical model (Figure 10). Along with the increase of the R_{MFR}, the temperatures also increased on both the hot and cold side of the sorption bed operating as a water–water heat exchanger. The lowest temperature was recorded by Probe 8 for all the analyzed R_{MFR} ratios, although CFD results were 2.0% ($R_{MFR} = 1.00$ and 1.33) or 2.1% ($R_{MFR} = 1.66$) higher in comparison to experimental results. The highest temperature was recorded by Probe A for all the analyzed R_{MFR} ratios. The relative difference between CFD and experimental results for Probe A ranged between 0.6% to 1.2%. The CFD and experimental results are qualitatively similar. The quantitative differences are presented in Figure 11 and range from 0.6% to 7.0%.

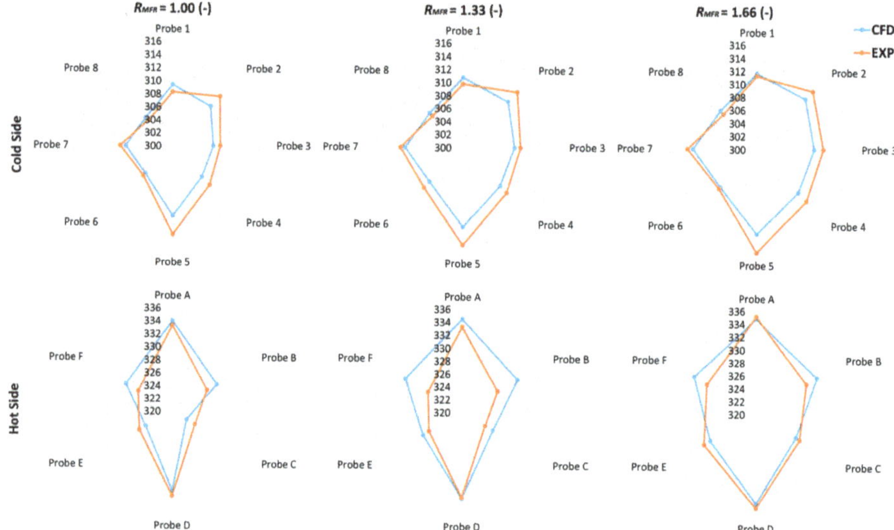

Figure 10. The temperatures in the multi-disc sorption bed operating as a water–water heat exchanger obtained within the experimental research and computational fluid dynamics (CFD) analysis.

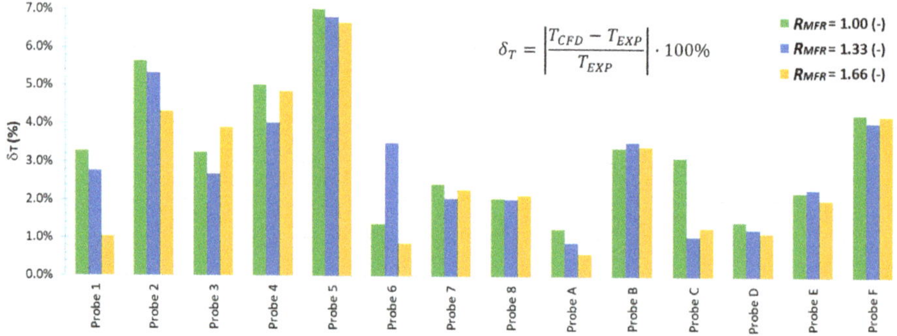

Figure 11. The relative difference between the temperatures in the multi-disc sorption bed operating as a water–water heat exchanger obtained in the experimental research and CFD analysis.

4.2. Temperature Field in the Sorption Bed

In the second stage of the research, the silica gel was placed into the disc-shaped packets in order to simulate the spatial distribution of temperature in the sorption bed at the end of the desorption cycle. The CFD analysis was performed for five d/a ratios, where d represents the inner diameter of the disc-shaped packet and a is the distance between the centers of the two adjacent packets.

The temperature along the line extending through the sorption bed from the water inlet (relative length = 0) to the water outlet (relative length = 1) is presented in Figure 12. Figures 13–17 present the temperature fields in the cross-sections of the sorption bed for all the analyzed *d/a* ratios (0.54, 0.62, 0.70, 0.78, and 0.86). The above-mentioned figures indicate that the temperature drop within the sorbent packets strongly depends on the *d/a* ratio. The lowest temperature in the center of the sorbent packet was approx. 312 K in the case of *d/a* = 0.54 and the highest temperature was approx. 325 K in the case of *d/a* = 0.86. The average sorbent temperature, as well as the temperature distribution in the bed, is highly influenced by the *d/a* ratio. The temperature distribution in the bed is the result of the superposition of the heat transferred from the heating water as well as the thermal characteristics of sorption kinetics. The proposed multi-disc sorption bed design is beneficial in terms of the spatial temperature distribution in the bed.

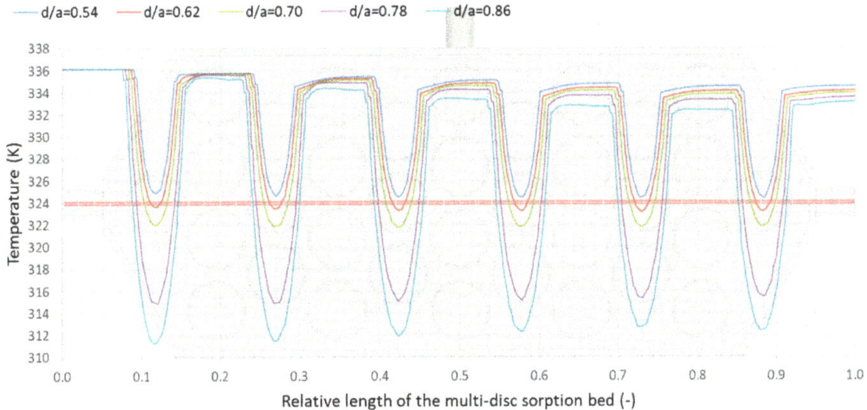

Figure 12. The temperature along the centerline of the sorption bed extending from the inlet (relative length = 0) to the outlet (relative length = 1).

The temperature drop within the sorbent entails the differences of the hot water temperature gradients (ΔT_{HW}) in the sorption bed presented in Table 4. The ΔT_{HW} is directly proportional to the heating power (*HP*) defined in Equation (11), which is one of the most important parameters of the adsorption chiller. Therefore, the increase in ΔT_{HW} obtained through the analyzed cases leads to a 69% increase in *HP* for constant mass flow rate ($\dot{m}_{HW}$) and specific heat (c_p) of heating water.

$$HP = \dot{m}_{HW} \cdot c_p \cdot \Delta T_{HW}, \tag{11}$$

On the other hand, the increased *d/a* ratio induces a higher pressure drop in the sorbent bed, which entails higher electricity consumption to power the hot water pump. This directly indicates the *d/a* ratio as the crucial parameter, which has to be considered in the design of the multi-disc sorption bed and properly balanced according to the specific requirements of the installation site.

Table 4. The hot water temperature (ΔT_{HW}) for all *d/a* ratios and the relative increase in heating power (HP%) as well as hot water pressure drop ($\Delta p_{HW\%}$) in relation to the *d/a* = 0.54.

d/a (-)	0.54	0.62	0.70	0.78	0.86
ΔT_{HW} (K)	1.71	2.10	2.24	2.52	2.89
$HP_\%$ (%)	0	23	31	47	69
$\Delta p_{HW\%}$ (%)	0	2	5	12	43

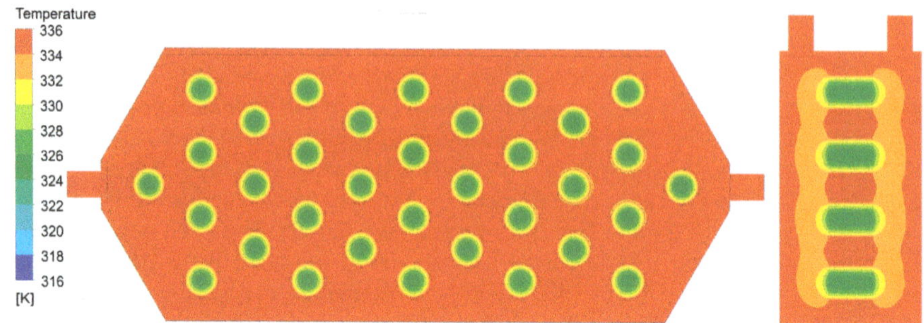

Figure 13. Temperature field in the cross-sections of the multi-disc sorption bed for $d/a = 0.54$.

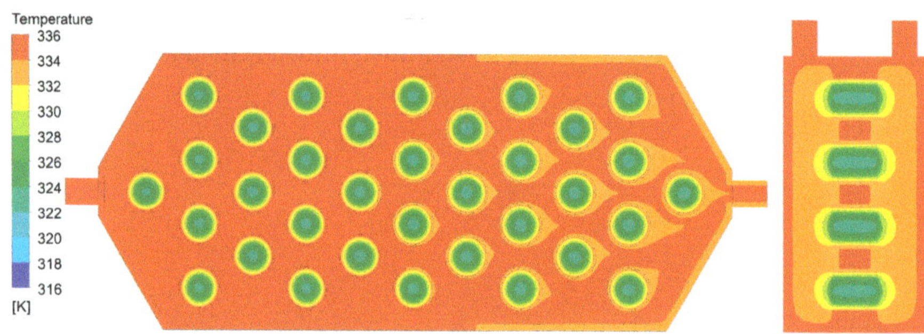

Figure 14. Temperature field in the cross-sections of the multi-disc sorption bed for $d/a = 0.62$.

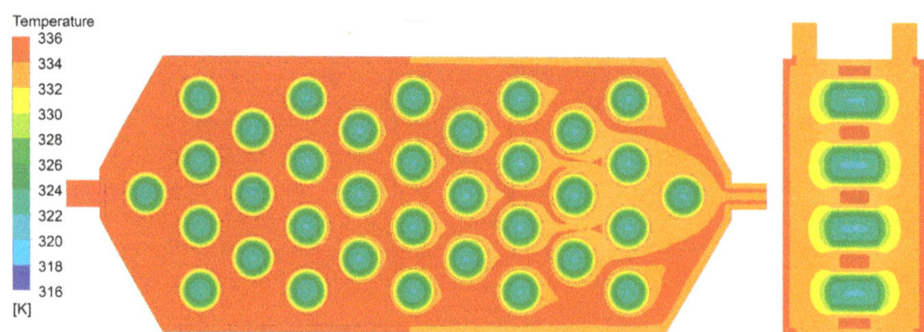

Figure 15. Temperature field in the cross-sections of the multi-disc sorption bed for $d/a = 0.70$.

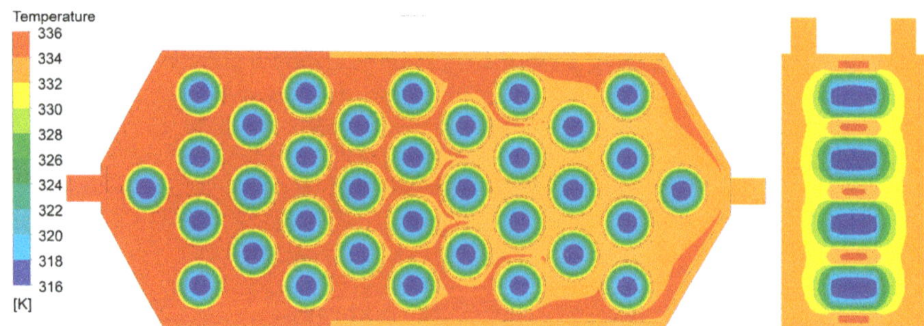

Figure 16. Temperature field in the cross-sections of the multi-disc sorption bed for $d/a = 0.78$.

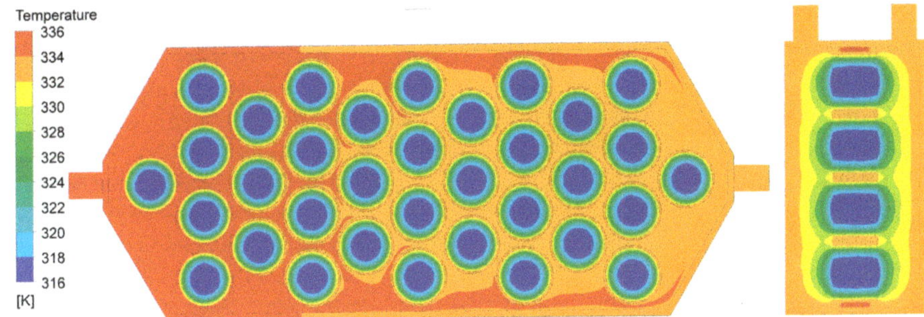

Figure 17. Temperature field in the cross-sections of the multi-disc sorption bed for $d/a = 0.86$.

The thickness of the copper wall of the disc-shaped packets was 1.5 mm for all the analyzed cases. The preliminary analysis revealed its insignificant influence on the temperature distribution in the bed. Such behavior results from the very high thermal conductivity of copper in comparison to the thermal conductivity of silica gel or water. Therefore, the wall thickness of the packets should be defined on the basis of mechanical analysis of the sorption bed rigidity while keeping in mind the requirement of minimizing both the bed mass and material cost.

4.3. Weight and Dimension Factors

The adsorption chiller performance is directly influenced by the thermal parameters, but the weight and dimensions of the sorption beds are, in some cases, the factors limiting the widespread application of the adsorption cooling technology. Therefore, the proper balance between the sorbent and metal in the sorption bed has to be thoroughly investigated.

One of the factors presenting the balance between the metal and the sorbent is the heat exchanger mass to sorbent ratio ($R_{HX/S}$) defined as the quotient of the total mass of the cylindrical heat transfer walls of the heat exchanger (m_{HX}) to the total sorbent mass (m_S). Lower values of $R_{HX/S}$ ratio are desirable in terms of the adsorption chiller performance and compact dimensions. Moreover, decreasing the fraction of heat exchanger mass in the total mass of the device leads to the growth of the COP because of supplying a greater portion of the thermal energy to the sorbent itself and not to the metal part of the sorption bed. Therefore, the d/a equal to 0.86 is the most advantageous design parameter value in the above context.

The ratio of heat transfer surface (S) to the sorbent mass (m_S) is convenient to assess the degree of a dynamic perfection of the sorption bed as it is proportional to the specific power; the larger the ratio the higher power per unit adsorbent mass can be obtained [32].

Both indicators for all the analyzed d/a ratios are depicted in Figure 18 and both decrease along with the increase of the d/a ratio.

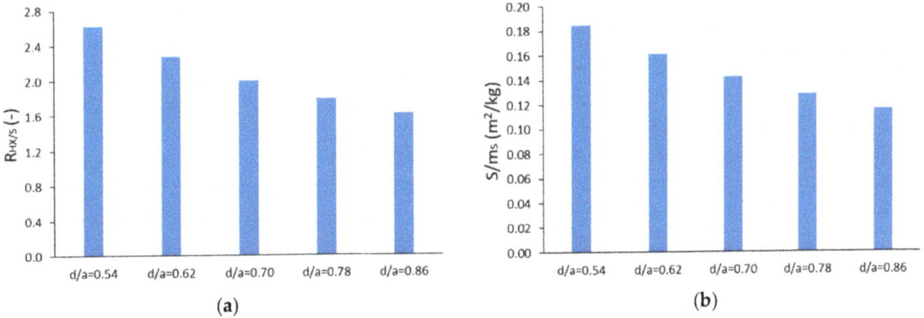

(a)　　　　(b)

Figure 18. The influence of multi-disc sorption bed design on $R_{HX/S}$ (**a**) and S/m_S (**b**) ratios.

5. Conclusions

The paper presents the analysis concerning the application of the multi-disc sorption bed of the adsorption chillers designed for chill and desalinated water production. The heat exchanger is the crucial element of the device because of its fundamental influence on the chiller performance indicators such as heating power (*HP*) or the coefficient of performance (*COP*). Moreover, it highly impacts the mass and dimensions of the sorption bed, and these factors are critical for the widespread utilization of the adsorption technology. The proposed multi-disc sorption bed contributes to the increase of heat transfer surface area and simultaneously assures a compact and lightweight design of the device. The developed design allows to place the device e.g., in the ceiling of the building or integrate it with solar panels, which will supply the device with the necessary heat. Such a solution allows to significantly expand the potential installation sites of the adsorption chillers and thus reduce one of the main disadvantages of these devices, which is the need to dedicate a large space for the installation of the adsorption chiller. Another advantage of the proposed construction is its potential for scalability consisting of adjusting the number of sorbent discs to the expected cooling capacity of the device or the possibility of installing two or more multi-disc sorption beds with sorbent packages one above another with the space between them being a vapor collector.

There is a great interest in the numerical modeling of adsorption chillers and particularly the sorption beds. However, the majority of the available models are characterized by significant limitations and only few of them are three-dimensional ones. Furthermore, some of the models presented in the literature are not experimentally validated and simulate only the packed bed not taking the heat exchanger geometry into consideration. Therefore, the developed and validated in-house algorithm allowing to model sorption along with thermal and flow processes in the 3D numerical model of the whole sorption bed filled the above-mentioned gap. The model takes into account the conjugate heat transfer as well as the thermal characteristics of the sorption phenomenon. In consequence, the developed model can contribute to the improvement of the design of adsorption beds by the capability of optimizing the sorption bed performance and geometry.

The CFD analysis performed with the use of the developed model incorporated into the commercial CFD code allowed to define the design of the investigated type of the sorption bed and its correlation with basic factors influencing the performance and dimensions of the sorption bed, such as gradient of heating water temperature (ΔT_{HW}), logarithmic mean temperature difference (*LMTD*), heat exchanger mass to sorbent ratio ($R_{HX/S}$), and heat transfer surface to sorbent mass ratio (S/m_s).

Funding: This research was funded by the National Science Centre of the Republic of Poland, grant number 2018/29/B/ST8/00442. The APC was funded by Ministry of Science and Higher Education of the Republic of Poland, grant number 944/P-DUN/2019.

Conflicts of Interest: The author declares no conflict of interest. The funders had no role in the design of the study; in the collection, analyses, or interpretation of data; in the writing of the manuscript, or in the decision to publish the results.

References

1. Alsaman, A.; Askalany, A.; Ahmed, M.; Ali, E.; Harby, K.; Diab, M. Simulation model for silica gel-water adsorption cooling system powered by renewable energy. In Proceedings of the 3rd International Conference on Energy Engineering Faculty, of Energy Engineering, Aswan University, Aswan, Egypt, 28–30 December 2015.
2. Elsheniti, M.B.; Hassab, M.A.; Attia, A.-E. Examination of effects of operating and geometric parameters on the performance of a two-bed adsorption chiller. *Appl. Therm. Eng.* **2018**, *146*, 674–687. [CrossRef]
3. Sultana, T. Effect of overall thermal conductance with different mass allocation on a two stage adsorption chiller employing re-heat scheme. Master's Thesis, Bangladesh University of Engineering and Technology, Dhaka, Bangladesh, 2008.

4. Khan, M.Z.I.; Alam, K.; Saha, B.B.; Hamamoto, Y.; Akisawa, A.; Kashiwagi, T. Parametric study of a two-stage adsorption chiller using re-heat—The effect of overall thermal conductance and adsorbent mass on system performance. *Int. J. Therm. Sci.* **2006**, *45*, 511–519. [CrossRef]

5. Saha, B.B.; Koyama, S.; Kashiwagi, T.; Akisawa, A.; Ng, K.C.; Chua, H.T. Waste heat driven dual-mode, multi-stage, multi-bed regenerative adsorption system. *Int. J. Refrig.* **2003**, *26*, 749–757. [CrossRef]

6. Sur, A.; Das, R.K. Review of technology used to improve heat and mass transfer characteristics of adsorption refrigeration system. *Int. J. Air Cond. Refrig.* **2016**, *24*, 1630003. [CrossRef]

7. Hassan, H.; Mohamad, A.; Alyousef, Y.; Al-Ansary, H. A review on the equations of state for the working pairs used in adsorption cooling systems. *Renew. Sustain. Energy Rev.* **2015**, *45*, 600–609. [CrossRef]

8. Voyiatzis, E.; Stefanakis, N.; Palyvos, J.; Papadopoulos, A. Computational study of a novel continuous solar adsorption chiller: Performance prediction and adsorbent selection. *Int. J. Energy Res.* **2007**, *31*, 931–946. [CrossRef]

9. Sztekler, K.; Kalawa, W.; Stefanski, S.; Krzywanski, J.; Grabowska, K.; Sosnowski, M.; Nowak, W. The influence of adsorption chillers on CHP power plants. *MATEC Web Conf.* **2018**, *240*, 05033. [CrossRef]

10. Sztekler, K.; Kalawa, W.; Nowak, W.; Stefanski, S.; Krzywanski, J.; Grabowska, K. Using the adsorption chillers for waste heat utilisation from the CCS installation. *EPJ Web Conf.* **2018**, *180*, 02106. [CrossRef]

11. Sztekler, K.; Kalawa, W.; Nowak, W.; Stefanski, S.; Krzywanski, J.; Grabowska, K. Using the adsorption chillers for utilisation of waste heat from rotary kilns. *EPJ Web Conf.* **2018**, *180*, 02105. [CrossRef]

12. Rezk, A.; Al-Dadah, R.; Mahmoud, S.; Elsayed, A. Effects of contact resistance and metal additives in finned-tube adsorbent beds on the performance of silica gel/water adsorption chiller. *Appl. Therm. Eng.* **2013**, *53*, 278–284. [CrossRef]

13. Saravanan, R.; Maiya, M.P. Thermodynamic comparison of water-based working fluid combinations for a vapour absorption refrigeration system. *Appl. Therm. Eng.* **1998**, *18*, 553–568. [CrossRef]

14. Kurniawan, A.; Rachmat, A. Others CFD Simulation of Silica Gel as an Adsorbent on Finned Tube Adsorbent Bed. *E3S Web Conf.* **2018**, *67*, 01014. [CrossRef]

15. Pyrka, P. Modelowanie trójzłożowej chłodziarki adsorpcyjnej. *Zesz. Energetyczne* **2014**, *1*, 205–216.

16. White, J. Literature review on adsorption cooling systems. *Lat. Am. Caribb. J. Eng. Educ.* 2013. Available online: https://www.researchgate.net/publication/289127089_LITERATURE_REVIEW_ON_ADSORPTION_COOLING_SYSTEMS (accessed on 8 December 2019).

17. Shahzad, M.W.; Ybyraiymkul, D.; Burhan, M.; Oh, S.J.; Ng, K.C. An innovative pressure swing adsorption cycle. *AIP Conf. Proc.* **2019**, *2062*, 020057.

18. Grabowska, K.; Sosnowski, M.; Krzywanski, J.; Sztekler, K.; Kalawa, W.; Zylka, A.; Nowak, W. The Numerical Comparison of Heat Transfer in a Coated and Fixed Bed of an Adsorption Chiller. *J. Therm. Sci.* **2018**, *27*, 421–426. [CrossRef]

19. Xu, S.Z.; Wang, L.W.; Wang, R.Z. Thermodynamic analysis of single-stage and multi-stage adsorption refrigeration cycles with activated carbon–ammonia working pair. *Energy Convers. Manag.* **2016**, *117*, 31–42. [CrossRef]

20. Starace, G.; Fiorentino, M.; Meleleo, B.; Risolo, C. The hybrid method applied to the plate-finned tube evaporator geometry. *Int. J. Refrig.* **2018**, *88*, 67–77. [CrossRef]

21. Fiorentino, M.; Starace, G. The design of countercurrent evaporative condensers with the hybrid method. *Appl. Therm. Eng.* **2018**, *130*, 889–898. [CrossRef]

22. Elsheniti, M.B.; Elsamni, O.A.; Al-dadah Raya, K.; Mahmoud, S.; Elsayed, E.; Saleh, K. Adsorption refrigeration technologies. In *Sustainable Air Conditioning Systems*; BoD—Books on Demand: Norderstedt, Germany, 2018.

23. Wu, W.-D.; Zhang, H.; Men, C. Performance of a modified zeolite 13X-water adsorptive cooling module powered by exhaust waste heat. *Int. J. Therm. Sci.* **2011**, *50*, 2042–2049. [CrossRef]

24. Sakoda, A.; Suzuki, M. Fundamental study on solar powered adsorption cooling system. *J. Chem. Eng. Jpn.* **1984**, *17*, 52–57. [CrossRef]

25. Sakoda, A.; Suzuki, M. Simultaneous Transport of Heat and Adsorbate in Closed Type Adsorption Cooling System Utilizing Solar Heat. *J. Sol. Energy Eng. Trans. Asme. J. Sol. Energy Eng.* **1986**, *108*, 239–245. [CrossRef]

26. Bahrehmand, H.; Khajehpour, M.; Bahrami, M. Finding optimal conductive additive content to enhance the performance of coated sorption beds: An experimental study. *Appl. Therm. Eng.* **2018**, *143*, 308–315. [CrossRef]

27. Kim, D.-S.; Chang, Y.-S.; Lee, D.-Y. Modelling of an adsorption chiller with adsorbent-coated heat exchangers: Feasibility of a polymer-water adsorption chiller. *Energy* **2018**, *164*, 1044–1061. [CrossRef]
28. Li, A.; Thu, K.; Ismail, A.B.; Shahzad, M.W.; Ng, K.C. Performance of adsorbent-embedded heat exchangers using binder-coating method. *Int. J. Heat Mass Transf.* **2016**, *92*, 149–157. [CrossRef]
29. Grabowska, K.; Krzywanski, J.; Nowak, W.; Wesolowska, M. Construction of an innovative adsorbent bed configuration in the adsorption chiller-Selection criteria for effective sorbent-glue pair. *Energy* **2018**, *151*, 317–323. [CrossRef]
30. Chang, K.-S.; Chen, M.-T.; Chung, T.-W. Effects of the thickness and particle size of silica gel on the heat and mass transfer performance of a silica gel-coated bed for air-conditioning adsorption systems. *Appl. Therm. Eng.* **2005**, *25*, 2330–2340. [CrossRef]
31. Grabowska, K.; Sosnowski, M.; Krzywanski, J.; Sztekler, K.; Kalawa, W.; Zylka, A.; Nowak, W. Analysis of heat transfer in a coated bed of an adsorption chiller. *MATEC Web Conf.* **2018**, *240*, 01010. [CrossRef]
32. Aristov, Y.I.; Glaznev, I.S.; Girnik, I.S. Optimization of adsorption dynamics in adsorptive chillers: Loose grains configuration. *Energy* **2012**, *46*, 484–492. [CrossRef]
33. Askalany, A.A.; Henninger, S.K.; Ghazy, M.; Saha, B.B. Effect of improving thermal conductivity of the adsorbent on performance of adsorption cooling system. *Appl. Therm. Eng.* **2017**, *110*, 695–702. [CrossRef]
34. Girnik, I.S.; Aristov, Y.I. Making adsorptive chillers more fast and efficient: The effect of bi-dispersed adsorbent bed. *Appl. Therm. Eng.* **2016**, *106*, 254–256. [CrossRef]
35. Demir, H.; Mobedi, M.; Ülkü, S. Effects of porosity on heat and mass transfer in a granular adsorbent bed. *Int. Commun. Heat Mass Transf.* **2009**, *36*, 372–377. [CrossRef]
36. Alam, K.C.A.; Saha, B.B.; Kang, Y.T.; Akisawa, A.; Kashiwagi, T. Heat exchanger design effect on the system performance of silica gel adsorption refrigeration systems. *Int. J. Heat Mass Transf.* **2000**, *43*, 4419–4431. [CrossRef]
37. Sosnowski, M.; Grabowska, K.; Krzywanski, J.; Nowak, W.; Sztekler, K.; Kalawa, W. The effect of heat exchanger geometry on adsorption chiller performance. *J. Phys. Conf. Ser.* **2018**, *1101*, 012037. [CrossRef]
38. Hong, S.; Ahn, S.; Kwon, O.; Chung, J. Optimization of a fin-tube type adsorption chiller by design of experiment. *Int. J. Refrig.* **2015**, *49*, 49–56. [CrossRef]
39. Ilis, G.G.; Demir, H.; Mobedi, M.; Saha, B.B. A new adsorbent bed design: Optimization of geometric parameters and metal additive for the performance improvement. *Appl. Therm. Eng.* **2019**, *162*, 114270. [CrossRef]
40. Gong, L.; Wang, R.; Xia, Z.; Chen, C. Design and performance prediction of a new generation adsorption chiller using composite adsorbent. *Energy Convers. Manag.* **2011**, *52*, 2345–2350. [CrossRef]
41. Rogala, Z. Adsorption chiller using flat-tube adsorbers—Performance assessment and optimization. *Appl. Therm. Eng.* **2017**, *121*, 431–442. [CrossRef]
42. Çağlar, A. The effect of fin design parameters on the heat transfer enhancement in the adsorbent bed of a thermal wave cycle. *Appl. Therm. Eng.* **2016**, *104*, 386–393. [CrossRef]
43. Pajdak, A.; Kudasik, M.; Skoczylas, N.; Wierzbicki, M.; Braga, L.T.P. Studies on the competitive sorption of CO2 and CH4 on hard coal. *Int. J. Greenh. Gas Control* **2019**, *90*, 102789. [CrossRef]
44. Pajdak, A.; Skoczylas, N.; Dębski, A.; Grzegorek, J.; Maziarz, W.; Kudasik, M. CO2 and CH4 sorption on carbon nanomaterials and coals—Comparative characteristics. *J. Nat. Gas Sci. Eng.* **2019**, *72*, 103003. [CrossRef]
45. Intini, M.; Goldsworthy, M.; White, S.; Joppolo, C.M. Experimental analysis and numerical modelling of an AQSOA zeolite desiccant wheel. *Appl. Therm. Eng.* **2015**, *80*, 20–30. [CrossRef]
46. Gurgel, J.; Andrade Filho, L.; Grenier, P.; Meunier, F. Thermal diffusivity and adsorption kinetics of silica-gel/water. *Adsorption* **2001**, *7*, 211–219. [CrossRef]
47. Demir, H.; Mobedi, M.; Ülkü, S. A review on adsorption heat pump: Problems and solutions. *Renew. Sustain. Energy Rev.* **2008**, *12*, 2381–2403. [CrossRef]
48. Krzywanski, J.; Grabowska, K.; Sosnowski, M.; Zylka, A.; Sztekler, K.; Kalawa, W.; Wójcik, T.; Nowak, W. Modeling of a re-heat two-stage adsorption chiller by AI approach. *MATEC Web Conf.* **2018**, *240*, 05014. [CrossRef]
49. Krzywanski, J.; Grabowska, K.; Herman, F.; Pyrka, P.; Sosnowski, M.; Prauzner, T.; Nowak, W. Optimization of a three-bed adsorption chiller by genetic algorithms and neural networks. *Energy Convers. Manag.* **2017**, *153*, 313–322. [CrossRef]

50. Krzywanski, J.; Grabowska, K.; Sosnowski, M.; Zylka, A.; Sztekler, K.; Kalawa, W.; Wojcik, T.; Nowak, W. An adaptive neuro-fuzzy model of a re-heat two-stage adsorption chiller. *Therm. Sci.* **2019**, *23*, 1053–1063. [CrossRef]

51. Papakokkinos, G.; Castro, J.; López, J.; Oliva, A. A generalized computational model for the simulation of adsorption packed bed reactors—Parametric study of five reactor geometries for cooling applications. *Appl. Energy* **2019**, *235*, 409–427. [CrossRef]

52. Sosnowski, M. Computational domain discretization in numerical analysis of forced convective heat transfer within packed beds of granular materials. *Eng. Mech.* **2018**, *2018*, 801–804.

53. Sosnowski, M.; Krzywanski, J.; Grabowska, K.; Gnatowska, R. Polyhedral meshing in numerical analysis of conjugate heat transfer. *EPJ Web Conf.* **2018**, *180*, 02096. [CrossRef]

54. Sosnowski, M. Computational domain discretization in numerical analysis of flow within granular materials. *EPJ Web Conf.* **2018**, *180*, 02095. [CrossRef]

55. Sosnowski, M.; Gnatowska, R.; Sobczyk, J.; Wodziak, W. Numerical modelling of flow field within a packed bed of granular material. *J. Phys. Conf. Ser.* **2018**, *1101*, 012036. [CrossRef]

56. Mitra, S.; Oh, S.T.; Saha, B.B.; Dutta, P.; Srinivasan, K. Simulation study of the adsorption dynamics of cylindrical silica gel particles. *Heat Transf. Res.* **2015**, *46*, 123–140. [CrossRef]

57. Khan, M.Z.I.; Alam, K.C.A.; Saha, B.B.; Akisawa, A.; Kashiwagi, T. Study on a re-heat two-stage adsorption chiller—The influence of thermal capacitance ratio, overall thermal conductance ratio and adsorbent mass on system performance. *Appl. Therm. Eng.* **2007**, *27*, 1677–1685. [CrossRef]

58. Wang, R.Z.; Xia, Z.Z.; Wang, L.W.; Lu, Z.S.; Li, S.L.; Li, T.X.; Wu, J.Y.; He, S. Heat transfer design in adsorption refrigeration systems for efficient use of low-grade thermal energy. *Energy* **2011**, *36*, 5425–5439. [CrossRef]

59. Antonellis, S.D.; Joppolo, C.M.; Molinaroli, L.; Pasini, A. Simulation and energy efficiency analysis of desiccant wheel systems for drying processes. *Energy* **2012**, *37*, 336–345. [CrossRef]

60. ANSYS. *Fluent Mosaic Technology Automatically Combines Disparate Meshes with Polyhedral Elements for Fast, Accurate Flow Resolution*; ANSYS: Canonsburg, PA, USA, 2018.

61. Sosnowski, M.; Gnatowska, R.; Sobczyk, J.; Wodziak, W. Computational domain discretization for CFD analysis of flow in a granular packed bed. *J. Theor. Appl. Mech.* **2019**, *57*, 833–842. [CrossRef]

62. Sosnowski, M.; Gnatowska, R.; Grabowska, K.; Krzywański, J.; Jamrozik, A. Numerical Analysis of Flow in Building Arrangement: Computational Domain Discretization. *Appl. Sci.* **2019**, *9*, 941. [CrossRef]

63. Eça, L.; Hoekstra, M. A procedure for the estimation of the numerical uncertainty of CFD calculations based on grid refinement studies. *J. Comput. Phys.* **2014**, *262*, 104–130. [CrossRef]

64. Sosnowski, M.; Krzywanski, J.; Scurek, R. A Fuzzy Logic Approach for the Reduction of Mesh-Induced Error in CFD Analysis: A Case Study of an Impinging Jet. *Entropy* **2019**, *21*, 1047. [CrossRef]

65. Celik, I.B.; Ghia, U.; Roache, P.J. Others Procedure for estimation and reporting of uncertainty due to discretization in CFD applications. *J. Fluids Eng. Trans. ASME* **2008**, *130*, 078001.

66. Kakac, S.; Liu, H.; Pramuanjaroenkij, A. *Heat Exchangers: Selection, Rating, and Thermal Design*; CRC Press: Boca Raton, FL, USA, 2012.

Article

Modelling Heat Pumps with Variable EER and COP in EnergyPlus: A Case Study Applied to Ground Source and Heat Recovery Heat Pump Systems

Antonella Priarone *, Federico Silenzi and Marco Fossa

DIME, University of Genova, Via all'Opera Pia 15a, 16145 Genova, Italy; federico.silenzi@edu.unige.it (F.S.); marco.fossa@unige.it (M.F.)

* Correspondence: a.priarone@unige.it

Received: 3 December 2019; Accepted: 5 February 2020; Published: 11 February 2020

Abstract: Dynamic energy modelling of buildings is a key factor for developing new strategies for energy management and consumption reduction. For this reason, the EnergyPlus software was used to model a near-zero energy building (Smart Energy Buildings, SEB) located in Savona, Italy. In particular, the focus of the present paper concerns the modeling of the ground source water-to-water heat pump (WHP) and the air-to-air heat pump (AHP) installed in the SEB building. To model the WHP in EnergyPlus, the Curve Fit Method was selected. Starting from manufacturer data, this model allows to estimate the *COP* of the HP for different temperature working conditions. The procedure was extended to the AHP. This unit is a part of the air-handling unit and it is working as a heat recovery system. The results obtained show that the HP performance in EnergyPlus can closely follow manufacturer data if proper input recasting is performed for EnergyPlus simulations. The present paper clarifies a long series of missed information on EnergyPlus reference sources and allows the huge amount of EnergyPlus users to properly and consciously run simulations, especially when unconventional heat pumps are present.

Keywords: heat pumps; EnergyPlus; buildings

---

1. Introduction

Energy savings and emissions reduction are two major keywords in the worldwide research scenario. One of the main responsible sectors is the buildings one, with approximately 40% of energy consumption and 36% of CO_2 emissions in the EU [1]. In this framework, it is easy to understand why energy dynamic simulations of buildings are of great importance and can be used to either develop innovative solutions for new buildings or to evaluate different retrofit interventions to enhance the energy performance of existing ones [2–4].

To decrease energy consumption and pollutant emissions, the first mandatory step is the reduction of building loads (i.e., building energy needs) by means of actions on the building envelope and related to the building operating conditions [5]. These include techniques to increase the external insulation combined with actions for the exploitation of the solar gains to reduce heating loads during winter [6].

After the activities devoted to minimizing the energy needs of the building, the second step is to select and correctly size innovative plants for heating and cooling that, if possible, include the use of renewable energies.

For example, solar energy can be exploited in thermal solar collectors to produce Domestic Hot Water (DHW) and in photovoltaic (PV) fields for powering the air conditioning system of buildings [7]. In particular, during the summer season, periods with higher solar radiation coincide with higher electrical energy demand for the cooling air conditioning systems and the use of PV modules helps to reduce the electrical national grid stress.

In recent years, one of the more frequently selected plant solutions for air conditioning in buildings has included reversible heat pumps (HPs) that allow to satisfy building requests both in heating and in cooling. Among them, Ground Coupled Heat Pumps (GCHP) are a very effective configuration, exploiting the near constant ground temperature during the year to increase the performance coefficients (*EER* in cooling mode and *COP* in heating mode) [8–10]. Performance of the ground-coupled heat pump greatly depends on several factors, including the fluid temperatures, the ground thermophysical properties and the configuration of Borehole Heat Exchangers (BHEs) used in the installation [11].

Another interesting plant solution is air-to-air heat pumps devoted to heat recovery on ventilation. This type of systems can be simple (i.e., direct expansion, DX) for the air conditioning of a single zone of a building, or more complex and sophisticated (i.e., included in air-handling unit, AHU), used for ventilation, humidification, filtration and air-conditioning with energy recovery (both active and passive techniques) of entire buildings.

It is apparent that one of the main problems in the correct sizing and in the modelling process of a heat pump is to take properly into account the variation of its performance in different operating conditions. In fact, the *COP* of a heat pump, even at full load, changes at different condensations and evaporation temperatures, which depend on the source and load side temperatures. For technological innovative solutions, the modelling process to include the heat pump in energy dynamic simulations can be difficult.

Lee et al. [12] recently developed a simplified heat exchanger model using artificial neural networks. Using a genetic algorithm, they managed to optimize the operating and design parameters of the heat exchanger in order to maximize the seasonal *EER* and the seasonal *COP* with respect to the outdoor temperature. Torregrosa-Jaime et al. [13] modelled the performances of a Variable Refrigerant Flow (VRF) equipment. They analysed the model proposed in EnergyPlus and they developed a new one using a BIM approach. Finally, they compared the results obtained with the manufacturer data.

Models for heat pumps pertain to two main groups, with two different approaches to the problem [14]. On one hand, there are the "equation fit models", which consider the heat pump as a black box, whose behavior is simulated by means of correlations with coefficients derived from manufacturer data. On the other hand, there are "deterministic models" that consider each component of the system applying energy and mass conservation equations.

The main differences between the two approaches are the amount of data requested and the application aim. The equation fit models are easier because they need only the knowledge of the performance at the operating conditions usually given by the manufacturer [15,16]. On the contrary, deterministic models also need data for specific HP components: these parameters often derive from dedicated measurement campaigns and are not provided by the manufacturer. This approach is useful for the study and design of specific components of the heat pump.

In dynamic simulations over long periods (e.g., yearly simulations for building response to environmental conditions and internal energy transfers), the working conditions of a heat pump change continuously, and it is mandatory to include, inside the model, at least the *COP* variation with temperature. The starting point are the data provided by the manufacturer in terms of the performance coefficients of the heat pump in heating and cooling at reference working conditions.

This paper deals with HP modelling in EnergyPlus environment. The application of the "equation fit model" is applied for modelling a water-to-water heat hump (Curve Fit Method [17]) and an air-to-air heat pump [18], the latter being applied for heat recovery purposes on air ventilation circuit.

To the above aim, a case study was taken into account and it refers to a n-ZEB building located at the Savona Campus of the University of Genova, Italy. In this case, a water-to-water heat pump is coupled with the ground and fed by water circulating in a borehole heat exchangers (BHEs) field. If the ground heat transfer is correctly evaluated, the returning fluid temperature (from the boreholes to the heat pump) is known with good approximation. The load side water temperature is imposed, based on the building request and on the operating condition of the distribution system (for the analyzed case composed by fancoils and radiators).

The investigated air-to-air heat pump is included in an air-handling unit and it serves as an energy recovery system on the exhaust air from ventilation. Thus, for that special air-to-air heat pump, the load side temperature is the external one, whereas the source side temperature is the return temperature from the building, nearly stable in both cooling and heating conditions.

By means of EnergyPlus simulations using "equation fit models", the variables *EER* and *COP* of both the heat pumps were evaluated for selected couples of source side and load side temperatures. For the water-to-water heat pump, the effect of the water volumetric flow rates (source and load side) was also taken into account by employing the manufacturer data related to the partial load factor (*PLF*) effect. For the air-to-air heat pump, the original contribution of the present study is the analysis of the suitable input datasets in case of unconventional heat pumps like the one with energy recovery here considered.

The good agreement between expected results and simulations validates the analysis. The present paper is in addition clarifying a long series of missed information on Energy Plus reference sources in order to allow the huge amount of Energy Plus users to efficiently, properly and consciously run simulations when considering temperature varying *COP*s.

2. Water-to-Water Heat Pump Model

This paragraph presents the literature models selected in the present study in order to properly address the input in the Energy Plus program to simulate water-to-water and air-to-air heat pumps (Figure 1) at temperature varying *COP*. The detailed description provided (and the related validations) here are original contributions of the present study, since Energy Plus references do not fully specify how the code can properly manage the running mode when inverse machines performance have to be customized in terms of manufacturer information.

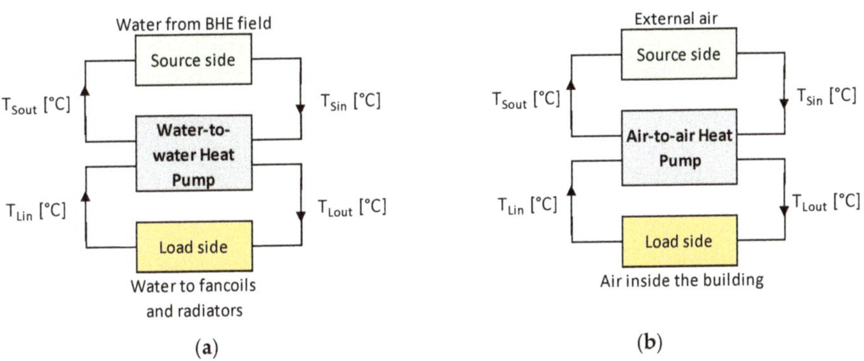

Figure 1. Scheme of HP operating conditions, (**a**) water-to-water HP, (**b**) air-to-air HP (without heat recovery).

In EnergyPlus, two different options are available to model the water-to-water heat pumps, i.e., the "Curve Fit Method" and the "Parameter estimation-based model" [15].

For the case study reported in this paper, the selected model is the "Curve Fit Method", which allows quicker simulation of the water-to-water heat pump, avoiding the drawbacks associated with the more computationally expensive "Parameter estimation-based model".

The variables that influence the water-to-water heat pump performance are mainly inlet water temperatures (source and load side) and water volumetric flow rates (source and load side).

The governing equations of the "Curve Fit Method" for the cooling and heating mode are the following ones [16]:

Cooling Mode:

$$\frac{\dot{Q}_C}{\dot{Q}_{C,ref}} = A_1 + A_2\left(\frac{T_{L,in}}{T_{ref}}\right) + A_3\left(\frac{T_{S,in}}{T_{ref}}\right) + A_4\left(\frac{\dot{V}_L}{\dot{V}_{L,ref}}\right) + A_5\left(\frac{\dot{V}_S}{\dot{V}_{S,ref}}\right) \tag{1}$$

$$\frac{P_C}{P_{C,ref}} = B_1 + B_2\left(\frac{T_{L,in}}{T_{ref}}\right) + B_3\left(\frac{T_{S,in}}{T_{ref}}\right) + B_4\left(\frac{\dot{V}_L}{\dot{V}_{L,ref}}\right) + B_5\left(\frac{\dot{V}_S}{\dot{V}_{S,ref}}\right) \tag{2}$$

Heating Mode:

$$\frac{\dot{Q}_H}{\dot{Q}_{H,ref}} = D_1 + D_2\left(\frac{T_{L,in}}{T_{ref}}\right) + D_3\left(\frac{T_{S,in}}{T_{ref}}\right) + D_4\left(\frac{\dot{V}_L}{\dot{V}_{L,ref}}\right) + D_5\left(\frac{\dot{V}_S}{\dot{V}_{S,ref}}\right) \tag{3}$$

$$\frac{P_H}{P_{H,ref}} = E_1 + E_2\left(\frac{T_{L,in}}{T_{ref}}\right) + E_3\left(\frac{T_{S,in}}{T_{ref}}\right) + E_4\left(\frac{\dot{V}_L}{\dot{V}_{L,ref}}\right) + E_5\left(\frac{\dot{V}_S}{\dot{V}_{S,ref}}\right) \tag{4}$$

where the parameters are defined as:

A_i, B_i, D_i, E_i	Equation fit coefficients for the cooling and heating mode [-].
T_{ref}	Reference temperature, 283.15 [K].
$T_{L,in}$	Load side inlet (in the HP) water temperature, [K].
$T_{S,in}$	Source side inlet (in the HP) water temperature, [K].
$\dot{V}_L$	Load side volumetric flow rate [m^3/s].
$\dot{V}_S$	Source side volumetric flow rate [m^3/s].
$\dot{Q}_C, \dot{Q}_H$	Load side heat transfer rate (cooling/heating mode) [W].
P_C, P_H	Power consumption (cooling/heating mode) [W].

The subscript "ref" indicates values at reference conditions that must be correctly specified. The reference temperature is always equal to 10 °C (283.15 K) and even when available data from manufacturer are provided at a different values, performance is to be recast to the above temperature.

In cooling mode, the reference conditions are when the heat pump operates at the highest (nominal) cooling capacity indicated in the manufacturer's technical references. The above condition does not match the real heat pump/chiller behavior since its performance can be even better than those at the nominal capacity, provided that the working temperature are "better" than the performance test ones. Similarly, in heating mode, the reference conditions are when the heat pump is operating at the highest (nominal) heating capacity.

In EnergyPlus, when selecting the "Curve Fit Method" to model water-to-water heat pumps, one must specify the parameters at the reference conditions and provide the equation fit coefficients.

Once the type of the water-to-water heat pump is selected, the generalized least square method is used for the evaluation of the coefficients A_i, B_i, D_i, E_i, based on the data available from the manufacturer's catalogue.

The performance coefficients (*EER* in cooling mode and *COP* in heating mode) are evaluated as the ratio between the useful heat transfer rate (load side) (Equations (1) and (3)) and the related power consumption (Equations (2) and (4)). Their equations as function of inlet temperatures and volumetric flow rates are, respectively:

Cooling Mode:

$$\frac{EER}{EER_{ref}} = \frac{A_1 + A_2\left(\frac{T_{L,in}}{T_{ref}}\right) + A_3\left(\frac{T_{S,in}}{T_{ref}}\right) + A_4\left(\frac{\dot{V}_L}{\dot{V}_{L,ref}}\right) + A_5\left(\frac{\dot{V}_S}{\dot{V}_{S,ref}}\right)}{B_1 + B_2\left(\frac{T_{L,in}}{T_{ref}}\right) + B_3\left(\frac{T_{S,in}}{T_{ref}}\right) + B_4\left(\frac{\dot{V}_L}{\dot{V}_{L,ref}}\right) + B_5\left(\frac{\dot{V}_S}{\dot{V}_{S,ref}}\right)} \tag{5}$$

Heating Mode:

$$\frac{COP}{COP_{ref}} = \frac{D_1 + D_2\left(\frac{T_{L,in}}{T_{ref}}\right) + D_3\left(\frac{T_{S,in}}{T_{ref}}\right) + D_4\left(\frac{\dot{V}_L}{\dot{V}_{L,ref}}\right) + D_5\left(\frac{\dot{V}_S}{\dot{V}_{S,ref}}\right)}{E_1 + E_2\left(\frac{T_{L,in}}{T_{ref}}\right) + E_3\left(\frac{T_{S,in}}{T_{ref}}\right) + E_4\left(\frac{\dot{V}_L}{\dot{V}_{L,ref}}\right) + E_5\left(\frac{\dot{V}_S}{\dot{V}_{S,ref}}\right)} \tag{6}$$

3. Air-to-Air Heat Pump Model

Air-to-air heat pump is here again modelled with an "equation fit model" [18].

Assuming constant supply air volumetric flow rate as operating conditions, the cooling and heating capacities and the *EER* and *COP* (and *EIR* = 1/*EER*) are only depending on temperatures and the selected equations to model the air-to-air heat pump are biquadratic ones. In particular, the performance depends on the "load air wet-bulb temperature" $T_{L,in\,wb}$ and the "source air dry-bulb temperature" $T_{S,in\,db}$ in cooling mode and on the "load air dry-bulb temperature" $T_{L,in\,db}$ and the "source air dry-bulb temperature" $T_{S,in\,db}$ in heating mode.

Cooling Mode:

$$\frac{\dot{Q}_C}{\dot{Q}_{C,ref}} = a_0 + a_1 \cdot T_{L,in\,wb} + a_2 \cdot T_{L,in\,wb}^2 + a_3 \cdot T_{S,in\,db} + a_4 \cdot T_{S,in\,db}^2 + a_5 \cdot T_{L,in\,wb} \cdot T_{S,in\,db} \tag{7}$$

$$\frac{EIR}{EIR_{ref}} = b_0 + b_1 \cdot T_{L,in\,wb} + b_2 \cdot T_{L,in\,wb}^2 + b_3 \cdot T_{S,in\,db} + b_4 \cdot T_{S,in\,db}^2 + b_5 \cdot T_{L,in\,wb} \cdot T_{S,in\,db} \tag{8}$$

Heating Mode:

$$\frac{\dot{Q}_H}{\dot{Q}_{H,ref}} = c_0 + c_1 \cdot T_{L,in\,db} + c_2 \cdot T_{L,in\,db}^2 + c_3 \cdot T_{S,in\,db} + c_4 \cdot T_{S,in\,db}^2 + c_5 \cdot T_{L,in\,db} \cdot T_{S,in\,db} \tag{9}$$

$$\frac{COP}{COP_{ref}} = d_0 + d_1 \cdot T_{L,in\,db} + d_2 \cdot T_{L,in\,db}^2 + d_3 \cdot T_{S,in\,db} + d_4 \cdot T_{S,in\,db}^2 + d_5 \cdot T_{L,in\,db} \cdot T_{S,in\,db} \tag{10}$$

In the previous equations, the parameters are defined as:

$\dot{Q}_C, \dot{Q}_H$	Load side heat transfer rate (cooling/heating mode) [W].
EER	Overall efficiency in cooling mode (thermodynamic circuit and fans) [-].
EIR	Performance coefficient in cooling mode (=1/*EER*) [-].
COP	Overall efficiency in heating mode (thermodynamic circuit and fans) [-].
a_i, b_i, c_i, d_i	Equation fit coefficients for the cooling and heating mode [-].
$T_{L,in\,wb}$	Load side inlet (in the HP) air wet bulb temperature, [K].
$T_{L,in\,db}$	Load side inlet (in the HP) air dry bulb temperature, [K].
$T_{S,in\,db}$	Source side air inlet (in the HP) dry bulb temperature, [K].

The subscript "ref" indicates values at reference conditions that must be correctly specified.

In EnergyPlus the reference conditions are required both in cooling and in heating mode. For the standard operating condition, in cooling mode, the reference load side air wet-bulb temperature $T_{L,in\,wb\,ref}$ is equal to 19.4 °C (with a corresponding reference load side air dry-bulb temperature $T_{L,in\,db\,ref}$ equal to 26.7 °C) whereas the source side air dry-bulb temperature is fixed at 35 °C. In heating mode, the reference load side air dry-bulb temperature $T_{L,in\,wb\,ref}$ is equal to 21.1 °C, whereas the source side air dry-bulb temperature is fixed at 8.3 °C.

In fact, for conventional reversible heat pumps, the load side conditions correspond to internal building ones (return air temperature T_{RA}) whereas source side conditions correspond to external ones (external air temperature T_{OA}).

4. The Case Study: The Smart Energy Building (SEB)

The Smart Energy Building (SEB) was conceived and built by the University of Genoa (Unige) as an innovative and high-performance building to meet goals of zero carbon emissions, energy and water efficiency and building automation. It is located in the Unige Campus of Savona, Italy.

The two storey building, in operation since February 2017, has a total floor area of about 1000 m^2. In particular, SEB is characterized by the presence of:

- high-performance thermal insulation materials for the envelope;
- ventilated facades;
- a photovoltaic field (21 kWp) on the roof;
- extremely low consumption led lamps;
- a rainwater collection system;
- a thermal system composed by:

✓ an air handling unit (AHU), associated to an air-to-air heat pump, installed on the roof, which performs functions such as circulating, cleaning and cooling/heating the air of the building;

✓ a ground coupled heat pump (GCHP), that produces cold/hot water to feed fancoils and radiators for cooling/heating purpose; the hot water is used during winter also for Domestic Hot Water (DHW) purposes;

✓ two solar thermal collectors, for DHW production purposes exclusively;

✓ an air source heat pump (ASHP), for DHW production as backup unit of the solar collectors.

The innovative nature of this building suggests the opportunity to analyze its performance from a dynamic point of view and to develop an energy model suitable for hourly simulations. EnergyPlus was selected to this aim. In particular, the present paper is focused on the modeling of the water-to-water ground coupled heat pump (GCHP) and on the air-to air heat pump associated to the AHU.

4.1. Modelling the Water-to-Water Ground Coupled Heat Pump (GCHP)

For the Smart Energy Building, the geothermal heat pump in operation is a Clivet brand, model WSHN-XEE2 MF 14.2, operating with brine (geothermal side) and water. In particular, data refer to operation with a mixture of water and propylene glycol at 30% on the source side.

The manufacturer catalogue provides the heat pump performance at full load as a function of source/load fluid temperatures. Table 1 represents the manufacturer data for the size 14.2, for the cooling mode.

The performance at full load related to the heating operating mode as a function of temperatures are provided by the Manufacturer in two different Tables depending on the range of the source side water temperature. For our test case, it is interesting to consider a wide range of working conditions for the source side temperature. In fact, for a GCHP with expected long life of operation, the temperature of the ground, starting from the undisturbed value, can change considerably in time [19] and consequently, the temperature of the fluid circulating in the BHE field changes.

The two manufacturer tables for heating mode differ for the selected values of the load side temperatures and thus, it is necessary to apply a proper interpolations. This is a typical problem in manufacturer data and cannot be managed in Energy Plus in a different way. The obtained combined dataset for heating mode is presented in Table 2. In grey are the data achieved by interpolation.

Table 1. Manufacturer data for reversible heat pump. The data are from Clivet manufacturer for HP model WSHN-XEE2 MF 14.2. Cooling Mode.

| Source Side Outlet Water Temperature, $T_{S,out}$ [°C] | Load Side Outlet Water Temperature, $T_{L,out}$ [°C] | | | | | | | | | | | | | | | | | |
| | 5 | | | 7 | | | 10 | | | 12 | | | 15 | | | 18 | | |
	kWt	kWe	EER	kWt	kWe	EER	kWt	kWe	EER	kWt	kWe	EER	kWt	kWe	EER	kWt	kWe	EER
25	41.7	7.67	5.43	44.3	7.75	5.72	48.1	7.89	6.10	50	8	6.25	54.6	8.2	6.66	58.9	8.51	6.92
30	39.9	8.67	4.60	42.5	8.76	4.86	46.1	8.89	5.19	48.1	9	5.34	52.4	9.18	5.71	56.6	9.46	5.98
35	38.1	9.67	3.93	40.7	9.76	4.17	44.1	9.89	4.46	46.2	10	4.62	50.1	10.16	4.93	54.3	10.4	5.22
40	35.4	10.92	3.24	38.3	10.9	3.50	41.6	11.1	3.75	43.5	11.1	3.92	47.3	11.3	4.19	51.3	11.6	4.42
45	32.9	12.3	2.68	35.3	12.4	2.86	38.2	12.5	3.06	40.2	12.6	3.19	43.6	12.7	3.43	47.2	13	3.63
50	29.8	13.8	2.16	32	13.9	2.30	34.8	14	2.49	36.5	14.1	2.59	39.9	14.3	2.79	43.1	14.5	2.97

Table 2. Manufacturer Data for reversible heat pump. The data are from Clivet manufacturer for HP model WSHN-XEE2. Heating Mode.

| Source Side Outlet Water Temperature, $T_{S,out}$ [°C] | Load Side Outlet Water Temperature, $T_{L,out}$ [°C] | | | | | | | | | | | |
| | 30 | | | 35 | | | 45 | | | 50 | | |
	kWt	kWe	COP	kWt	kWe	COP	kWt	kWe	COP	kWt	kWe	COP
0	41.8	7.37	5.67	41.3	8.35	4.95	40.5	10.7	3.79	39.4	12.2	3.23
1	43.1	7.39	5.83	42.4	8.37	5.07	41.6	10.7	3.89	40.4	12.2	3.32
3	45.5	7.44	6.12	44.9	8.43	5.33	43.8	10.7	4.10	42.6	12.2	3.50
5	46.5	7.47	6.22	46.3	8.48	5.45	45.4	10.7	4.24	43.8	12.2	3.59
7	49.4	7.53	6.55	49.0	8.54	5.73	48.0	10.7	4.48	46.25	12.2	3.79
10	53.7	7.62	7.04	53.2	8.65	6.15	51.8	10.8	4.79	49.85	12.25	4.07
12	56.7	7.70	7.36	56.2	8.73	6.44	54.7	10.9	5.02	52.55	12.35	4.26
15	61.7	7.83	7.88	60.9	8.86	6.87	59.2	11.0	5.38	56.75	12.4	4.58
17	65.2	7.92	8.24	64.4	8.96	7.19	62.5	11.0	5.68	59.85	12.45	4.81

It is important to notice that the performances in Tables 1 and 2 are provided as a function of the outlet temperatures $T_{S,out}$ and $T_{L,out}$ whereas Equations (11)–(16) contain the inlet ones, $T_{S,in}$ and $T_{L,in}$.

However, the manufacture catalogue provides details about the operating conditions related to the performances of Tables 1 and 2. In detail, the *EER* and *COP* data refer to following imposed temperature difference at the load and source sides:

Cooling (Table 1):

$$T_{L,in} = T_{L,out} + 5\,^\circ C \qquad T_{S,in} = T_{S,out} - 5\,^\circ C \tag{11}$$

Heating (Table 2):

$$T_{L,in} = T_{L,out} - 5\,^\circ C \qquad T_{S,in} = T_{S,out} + 5\,^\circ C \quad \text{for } T_{S,out} = 0, 1, 3\,^\circ C$$
$$T_{S,in} = T_{S,out} + 3\,^\circ C \quad \text{for } T_{S,out} = 5, 7, 10, 12, 15, 17\,^\circ C \tag{12}$$

For a complete analysis, it is necessary to take into account also the effect of water volumetric flow rates (source and load side) on the heat pump performance. The manufacturer provides only little information about the effect of the partial load factor *PLF* on the *EER* and *COP* of the water-to-water heat pump and in particular the performance at *PLF* of 67% and 33%. Both the *EER* and the *COP* are enhanced at partial load, according to Table 3.

Table 3. Manufacturer data for Clivet model WSHN-XEE2. Effect of *PLF* on the HP performance.

PLF	*EER/EER*$_{full\ load}$	*COP/COP*$_{full\ load}$
0.33	1.080	1.146
0.67	1.032	1.103
1	1.000	1.000

Considering that both the source and load sides of the HP work at constant temperature difference according to Equations (11) and (12), the *PLF* represents not only the ratio between actual cooling or heating capacity and the maximum value but also the corresponding ratio between the water volumetric flow rates at load side. From the values of *EER* or *COP* in Table 3 it is possible to deduce the power consumption (cooling and heating mode) and the source side heat transfer rate and, as a consequence, the water volumetric flow rates at source side.

The coefficients A_i, B_i, D_i and E_i of Equations (1)–(6) are not available from manufacturer references. The only way for accessing them is to iteratively guess their correct value by comparison with the available datasheet values and by minimizing an error. In this paper, a simple optimum search process was applied to cooling or heating capacity and power consumption values provided in the manufacturer catalogue.

The final calculated coefficients are presented in Table 4.

Table 4. Calculated coefficients for the "Curve Fit Method" for the water-to-water HP.

A_1	0.957	D_1	0.088
A_2	0.407	D_2	−0.090
A_3	−1.326	D_3	0.012
A_4	0.076	D_4	0.992
A_5	0.916	D_5	0.001
B_1	−5.181	E_1	1.100
B_2	−1.927	E_2	8.056
B_3	6.627	E_3	−10.091
B_4	−1.503	E_4	1.862
B_5	2.930	E_5	0.089

Figures 2 and 3 compare the manufacturer data with the values obtained with the correlations 1–4 using the optimum calculated coefficients of Table 4. In particular, the graphs show the cooling/heating capacities and the power consumptions for cooling and heating, respectively, as a function of the source side outlet water temperature $T_{S,out}$ with the load side outlet water temperature $T_{L,out}$ as a parameter and considering the three conditions of load, namely $PLF = 1, 0.67, 0.33$.

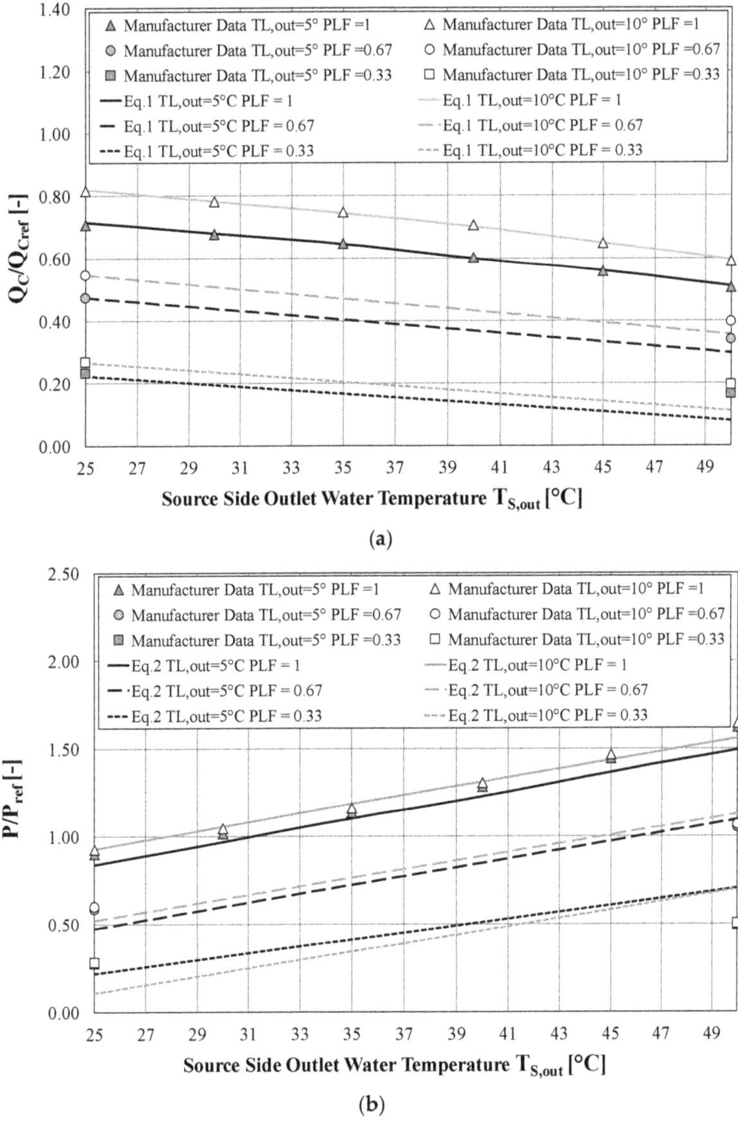

Figure 2. (a) $Q_C/Q_{C\,ref}$ and (b) P/P_{ref} comparison for cooling mode.

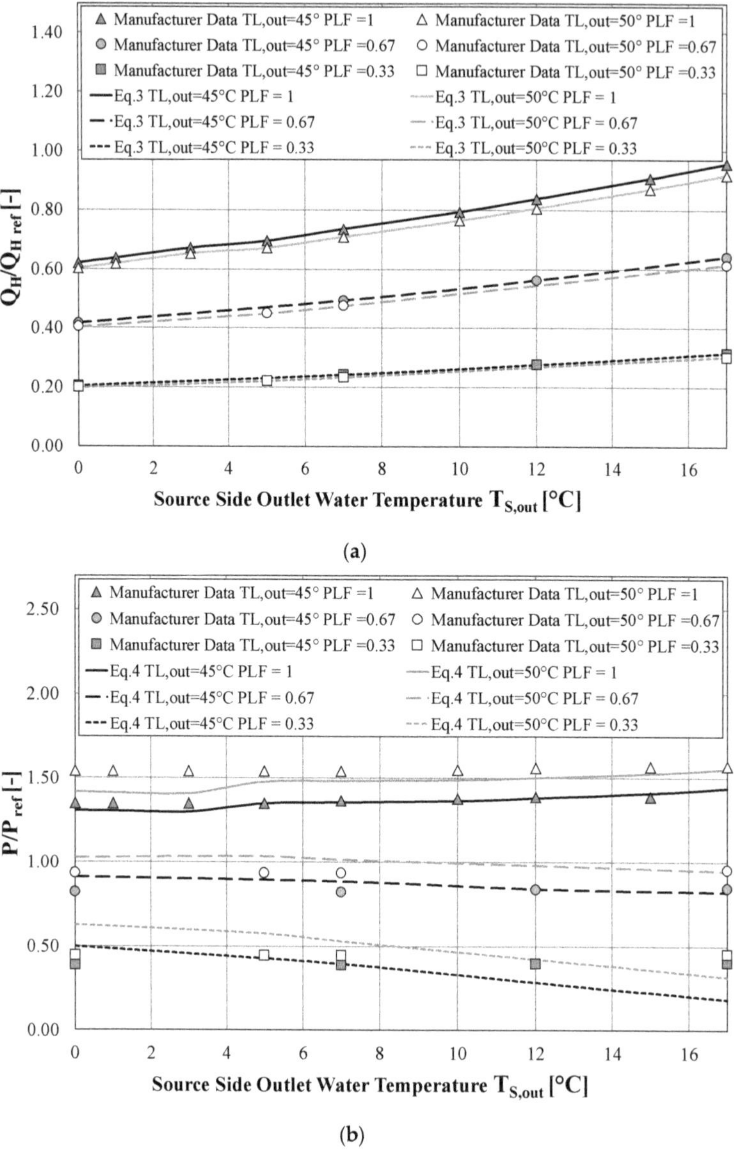

Figure 3. (**a**) $Q_H/Q_{H\,ref}$ and (**b**) $P/P_{\,ref}$ comparison for heating mode.

During the summer season, the cooling capacity $\dot{Q}_C$ decreases to increase the source side outlet water temperature $T_{S,out}$ (fluid temperature entering in the BHE field) and increases to increase the load side outlet water temperature $T_{L,out}$ (fluid temperature to fancoils and radiators). On the contrary, the power consumption P increases to increase the source side outlet water temperature $T_{S,out}$ whereas the effect of the load side outlet water temperature $T_{L,out}$ is nearly negligible. As expected, both the cooling capacity and the power consumption decreased by decreasing the partial load factor PLF.

During winter, on the other hand, the heating capacity $\dot{Q}_H$ increases for increasing source side outlet water temperature $T_{S,out}$ and slightly decreases for increasing load side outlet water temperature

$T_{L,out}$. The power consumption P is marginally affected by the source side outlet water temperature $T_{S,out}$, whereas it increases with the load side outlet water temperature $T_{L,out}$.

The particular trend of the curve given by Equations (3) and (4), with two slight inflection points for $T_{S,out} = 3$ and $5\,^\circ\text{C}$, is due to the particular operating conditions for the manufacture catalogue in heating mode. In fact, manufacture tables in the heating mode are built for different imposed temperature differences at the load and source sides, according to Equation (12). Thus, at different source sides, "outlet" water temperatures $T_{S,out}$ correspond the same source side "inlet" water temperatures $T_{S,in} = 8\,^\circ\text{C}$ that represents the input of Equations (3) and (4).

As for the cooling case, both the heating capacity and the power consumption decrease by decreasing the partial load factor *PLF*.

The agreement between manufacture dataset and "equation fit models" approach is good, with an average relative error between less than 7%, for both cooling and heating mode at full load and lower than 15% considering also the *PLF* = 0.67 and 0.33.

4.2. Modelling the Air-to-Air Heat Pump

For the Smart Energy Building, the selected air-to-air heat pump associated to the air handling unit (AHU) is the Clivet Zephir CPAN-XHE3 Size 3, with a standard air flow of 4600 m³/h. This volumetric flow rate fulfils the ventilation requested by the Italian standards for the SEB building in terms of its volume and expected occupancy levels.

This air unit is very peculiar, especially if compared with the options conceived and available in Energy Plus. This system is a primary-air plant with a thermodynamic recovery of the energy contained in the return air. The primary air comes entirely from outdoor (fresh-air) at temperature T_{OA} whereas the return-air comes from the building inner rooms at temperature T_{RA}. Return air, before being released to the atmosphere, exchanges heat with the condenser in cooling mode and with the evaporator in heating mode. Return-air represents a favorable thermal source stable in time, offering lower temperature on the condenser side in cooling mode and higher temperature on the evaporator side in heating mode. As a consequent, the energy required by the compressors is reduced up to 50% [20].

The manufacturer catalogue provides the reversible heat pump performances as a function of external air temperature T_{OA} (dry bulb/wet bulb) and supply air temperature T_{SA}. Moreover, the manufacturer catalogue reports two different types of performance coefficients, the thermodynamic efficiencies (*EER*$_{\text{th}}$ and *COP*$_{\text{th}}$) and the overall efficiencies (*EER* and *COP*) that consider also the power of the auxiliary systems.

In cooling mode, the selected supply humidity ratio is equal to 11 g$_{\text{vap}}$/kg$_{\text{air}}$ and the reference return air temperature T_{RA} is 26 °C. In heating mode, the reference return air temperature T_{RA} is 20/12 °C (dry bulb/wet bulb). To model the air-to-air HP in EnergyPlus, the data corresponding to the "MC" operation mode were not considered, that imply post-heating equal to zero in cooling mode.

The distinctive operating conditions of the present heat pump (with energy recovery) allow it to reach high values of performance coefficients but create some challenges in modelling the system in EnergyPlus. In fact, the "load side" temperature becomes the external air temperature T_{OA} whereas the "source side" temperature is the return air temperature T_{RA} both in cooling and in heating modes. Consequently, the reference conditions suggested from EnergyPlus (Par. 2.2) are no longer valid and new reference conditions are defined for the analyzed present heat pump.

In particular, in cooling mode, the new reference external air temperature T_{OA} is set to 40/25 °C (dry-bulb/wet-bulb) whereas the reference return air temperature T_{RA} is set to 26 °C (Table 5). In heating mode, the new reference external air temperature T_{OA} is set to −5 °C (dry-bulb) whereas the reference return air temperature T_{RA} is set to 20/12 °C (dry bulb/wet bulb) (Table 6).

Table 5. Datasheet values in cooling mode for present study analyses (manufacturer data in gray). Air handling unit model Zephir CPAN-XHE3 (air flow 4600 m^3/h, supply humidity ratio 11 g$_{vap}$/kg$_{air}$).

Reference Conditions					
Ref. External Air Temperature (Dry-Bulb) T_{OAdb} [°C]	**Ref. External Air Temperature (Wet-Bulb)** T_{OAwb} [°C]	**Ref. Return Air Temperature (Dry-Bulb)** T_{RAdb} [°C]	**Ref. Cooling Capacity [W]**	**Ref. Compressor + Fan Power [W]**	**Ref. *EER* [-]**
40	25	26	41,900	16,115	2.6

Performance Data					
External AIR Temperature (Dry-Bulb) T_{OAdb} [°C]	**External Air Temperature (Wet-Bulb)** T_{OAwb} [°C]	**Return Air Temperature (Dry-Bulb)** T_{RAdb} [°C]	**Cooling Capacity [W]**	**Compressor + Fan Power [W]**	***EER* [-]**
40	25	26	41,900	16,115	2.60
35	24	26	38,700	13,345	2.90
32	23	26	34,000	10,000	3.40
30	22	26	29,100	6929	4.20
28	21	26	23,600	4917	4.80
25	19	26	8100	2132	3.80
40	25	22	41,900	14,249	2.94
35	24	22	38,700	12,009	3.22
32	23	22	34,000	8794	3.87
30	22	22	29,100	6095	4.77
28	21	22	23,600	4292	5.50
25	19	22	8100	1735	4.67
40	25	20	41,900	13,383	3.13
35	24	20	38,700	11,273	3.43
32	23	20	34,000	8224	4.13
30	22	20	29,100	5675	5.13
28	21	20	23,600	3971	5.94
25	19	20	8100	1515	5.35

Table 6. Datasheet values in heating mode for present study analyses (manufacturer data in gray). Air handling unit model Zephir CPAN-XHE3 (air flow 4600 m^3/h).

Reference Conditions				
Ref. External Air Temperature (Dry-Bulb) T_{OAdb} [°C]	Ref. Return Air Temperature (Dry-Bulb) T_{RAdb} [°C]	Ref. Heating Capacity [W]	Ref. Compressor + Fan Power [W]	Ref. COP [-]
−5	20	49,700	11,044	4.50
Performance Data				
External Air Temperature (Dry-Bulb) T_{OAdb} [°C]	Return Air Temperature (Dry-Bulb) T_{RAdb} [°C]	Heating Capacity [W]	Compressor + fan power [W]	COP [-]
−5	20	49,700	11,044	4.50
0	20	49,500	12,375	4.00
2	20	46,200	11,268	4.10
7	20	37,100	8065	4.60
12	20	28,400	5462	5.20
−5	22	49,700	10,592	4.69
0	22	49,500	11,717	4.22
2	22	46,200	10,738	4.30
7	22	37,100	7712	4.81
12	22	28,400	5254	5.41
−5	26	49,700	9469	5.25
0	26	49,500	10,633	4.66
2	26	46,200	9703	4.76
7	26	37,100	6868	5.40
12	26	28,400	4605	6.17

The EnergyPlus model for the air-to-air heat pump implements Equations (7)–(10) that express the cooling/heating capacities $\dot{Q}_C$, $\dot{Q}_H$ and the *EIR*, *COP* as a function of both the external air temperature T_{OA} (dry-bulb/wet-bulb) and the return air temperature T_{RA}. Unfortunately (again a typical case), the data provided by the manufacturer are a function of a unique value of the return temperature T_{RA}, namely 26 °C in cooling and 20/12 °C (dry bulb/wet bulb) in heating.

Thus, it is necessary to create an extended database to obtain, by optimization, the coefficients a_i, b_i, c_i and d_i of Equations (7)–(10). The selected return temperatures T_{RA} to extend the dataset are 20, 22, 26 °C.

By keeping constant the air volumetric flow rate, for the same external and supply conditions (temperature and humidity), also the cooling and heating capacities remain constant. On the contrary, modifying the return temperature conditions changes the "source temperature" and as a consequence, the performance coefficients (*EER* and *COP*) and the compressor power are modified.

The values of the thermodynamic performance coefficients (*EER*$_{th}$ and *COP*$_{th}$) for the new values of the return temperatures T_{RA} are obtained by multiplying the corresponding Carnot performance coefficients (*EER*$_{Carnot}$ and *COP*$_{Carnot}$) based on the evaporator and condenser temperatures, by two sets of constants C_{Ci} and C_{Hi}.

The coefficients C_{Ci} and C_{Hi} are calculated here from Carnot law and manufacturer data according to the expressions below. Moreover, they are assumed to be dependent on the supply air temperature T_{SA} but independent of the return temperatures T_{RA}.

$$EER_{th} = C_{Ci} \cdot EER_{Carnot} = C_{Ci} \cdot \frac{T_{evap}}{T_{cond} - T_{evap}} \tag{13}$$

$$COP_{th} = C_{Hi} \cdot COP_{Carnot} = C_{Hi} \cdot \frac{T_{cond}}{T_{cond} - T_{evap}} \tag{14}$$

The evaporator temperature T_{evap} is assumed to be nearly equal to the supply air temperature T_{SA} whereas the condenser temperature T_{cond} is evaluated by means of energy balances on the components of the HP.

The condenser temperature T_{cond} is assumed to be nearly equal to the supply air temperature T_{SA} whereas the evaporator temperature T_{evap} is evaluated by means of energy balances on the components of the HP.

Finally, overall efficiencies (*EER* and *COP*) are deduced by assuming the fan power consumption as constant and equal to 1 kW for all the operating conditions considered.

The results of this analysis are presented in Tables 5 and 6, in cooling and heating mode respectively (data in grey, original data provided by the manufacturer in white).

Finally, by means of an optimum search process comparing the performance values of Tables 5 and 6, the coefficients a_i, b_i, c_i and d_i of Equations (7)–(10) were obtained and the results are presented in Table 7.

Table 7. Calculated coefficients for the "Equation Fit Approach", air-to-air heat pump.

a_0	−6.04980	b_0	−2.20000	c_0	−0.06076	d_0	0.59269
a_1	0.48670	b_1	0.11000	c_1	−0.00423	d_1	0.02513
a_2	−0.00820	b_2	0.00000	c_2	−0.00148	d_2	−0.00190
a_3	0.00000	b_3	0.00300	c_3	0.08791	d_3	0.05807
a_4	0.00000	b_4	0.00056	c_4	−0.00186	d_4	−0.00171
a_5	0.00000	b_5	0.00000	c_5	−0.00053	d_5	−0.00094

As an example, Figures 4 and 5 show the cooling and heating capacities and the HP performances (*EIR* and *COP*) as a function of external conditions T_{OA} and return temperature T_{RA} as parameter. In particular, the manufactured data reported in Tables 5 and 6 are compared with the curves obtained by using Equations (7)–(10) with the least square error coefficients of Table 7.

As expected, not changing the volumetric flow rate and the external T_{OA} and supply conditions (temperature and humidity), the cooling and heating capacities $\dot{Q}_C$, $\dot{Q}_H$ remain almost constant for the different return air conditions TRA (Figures 4a and 5a). The cooling capacity $\dot{Q}_C$ (requested by the building) increases with the external temperature T_{OA} whereas the heating capacity $\dot{Q}_H$ (requested by the building) decreases by increasing the external temperature T_{OA}.

On the contrary, the performance parameter *EIR* (=1/*EER*) and *COP* depend on both the external and the return air temperature (Figures 4b and 5b). In cooling mode, the *EIR* increases with the external air temperature T_{OA} (load side temperature) and increases with the return air temperature T_{RA} (source side temperature). In heating mode, the *COP* decreases as the return air temperature T_{RA} is increased (source side temperature) whereas it decreases with the external air temperature (load side temperature) for $T_{OA} > 0\,°C$. For $T_{OA} < 0\,°C$, the *COP* increases with the external air temperature because of the energy consumption of the defrost contribution.

The agreement between manufacturer data and best-fit curves is good and the coefficients can be implemented in EnergyPlus to represent the behaviour of the present air-to-air heat pump. The average relative error (fit profiles vs. manufactured data) in cooling is about 2.3% for the cooling capacity $\dot{Q}_C$ and 3.3% for the *EER*. In heating mode, the average relative error is 2.4% for the heating capacity $\dot{Q}_H$ and 2.6% for the *COP*.

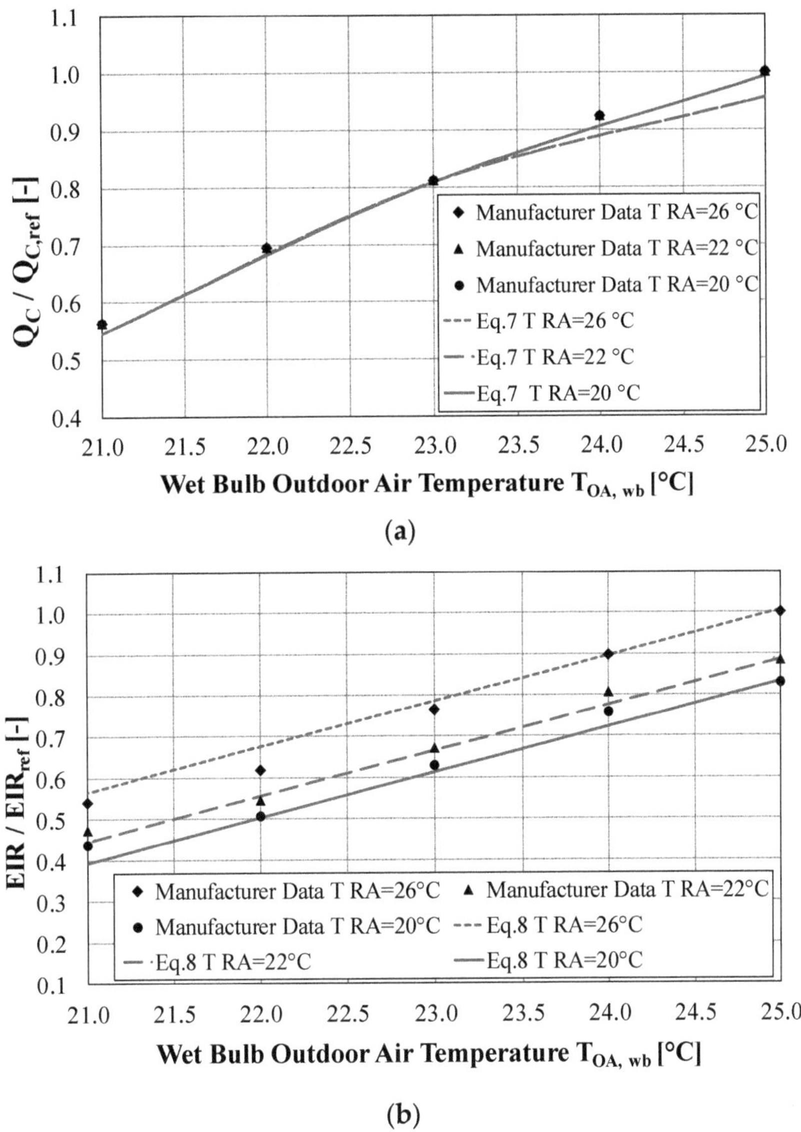

Figure 4. Cooling capacity (a) and HP performance (b) in cooling mode: comparison between Table 5 and Equations (7) and (8) with Table 7 coefficients.

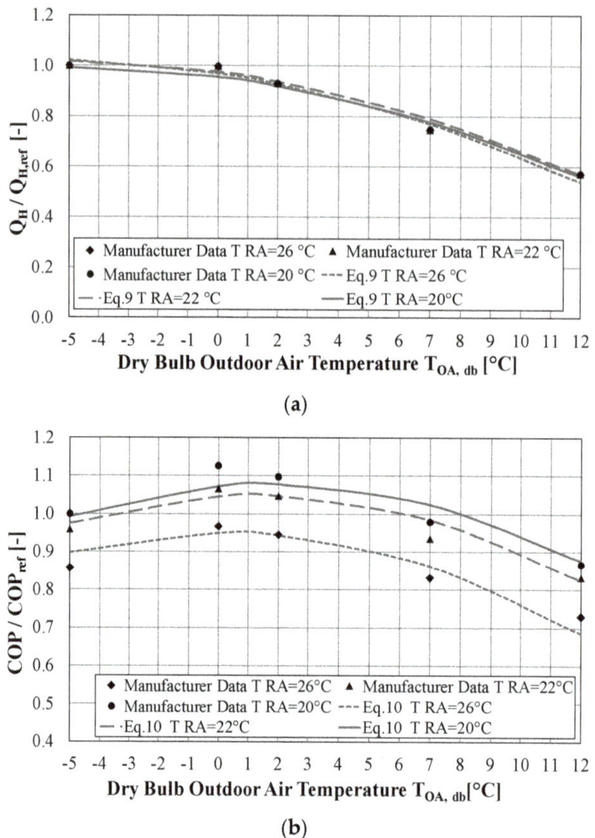

(a)

(b)

Figure 5. Heating capacity (**a**) and HP performance (**b**) in heating mode: comparison between Table 6 and Equations (9) and (10) with Table 7 coefficients.

5. Results

5.1. Water-to-Water Heat Pump

The proposed "Curve Fit Method" presented in the previous paragraphs was validated with reference benchmark simulations in EnergyPlus.

A simplified model was created for this purpose, with a building able to work at nearly constant operating conditions for the whole simulation period, i.e., 1 month. The modelled building is equipped with the GCHP Clivet WSHN-XEE2 MF 14.2 and both cooling and heating modes are simulated. Different working conditions are analysed, imposing different load side outlet $T_{L,out}$ and source side inlet $T_{S,in}$ water temperature. The load of the building and the thermal response of the ground are properly calibrated to maintain the desired temperature difference at the source and load sides (Equations (11) and (12)).

The results for the full load operating conditions are presented in Tables A1 and A2 in Appendix A where the first two columns represent the imposed operating temperatures. From EnergyPlus simulations, it is possible to infer the inlet load and outlet source temperatures and verify that the temperature differences at the source and load sides are comparable to the desired values (Equations (11) and (12)). The performance values (*EER* and *COP*) are evaluated from the ratio between the simulated values of cooling or heating capacity $\dot{Q}_C, \dot{Q}_H$ and the electrical consumptions P. These simulated performances are then compared with the values calculated by means of the Curve Fit Method with a very good agreement.

The average absolute relative error on *EER* in cooling is nearly 0.15% whereas *COP* in heating is nearly 0.6%. The comparison, in terms of *EER* and *COP*, is represented graphically in Figures 6 and 7 in cooling and heating mode, respectively.

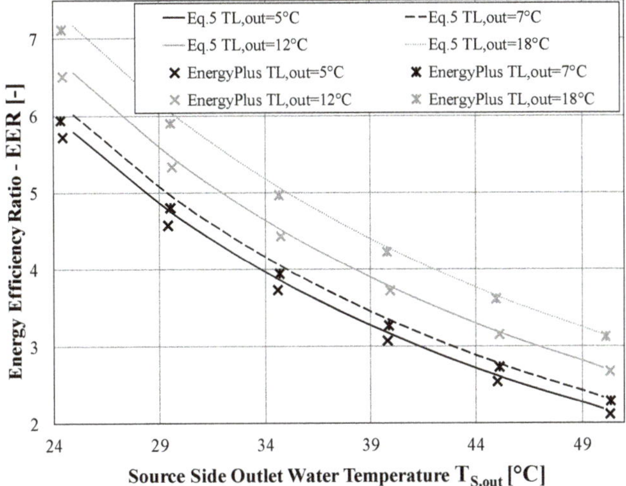

Figure 6. *EER* in cooling mode: comparison between EnergyPlus simulations and equation fit model approach.

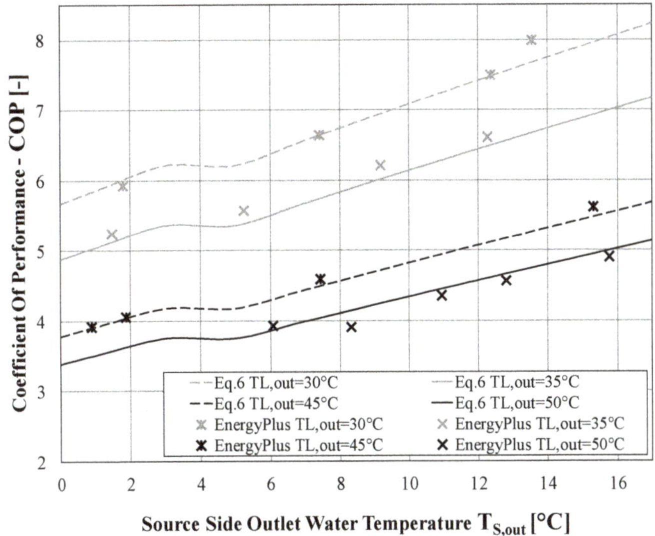

Figure 7. *COP* in heating mode: comparison between EnergyPlus simulations and equation fit model approach.

5.2. Air-to-Air Heat Pump

The equation fit model approach was implemented in EnergyPlus also for the air-to-air heat pump, by means of Equations (7)–(10) with the coefficients in Table 7.

Similarly, in this case, a simplified building model was created with nearly constant operating conditions for the whole simulation duration, i.e., 1 month. The modelled building is equipped with the Clivet Zephir CPAN-XHE3 Size 3 and both cooling and heating are simulated.

Different operating conditions were simulated, namely the ones presented in Tables A3 and A4 in Appendix A for cooling and heating mode, respectively. In the tables, the results of EnergyPlus simulations are reported and compared with the data obtained with the implemented equation fit model. The agreement is very good, with an average relative error of almost 1.5% for the cooling capacity $\dot{Q}_C$, 1.6% for the *EER*, 5.1% for the heating capacity $\dot{Q}_H$ and 0.26% for the *COP*. Figures 8 and 9 show graphically the same comparison.

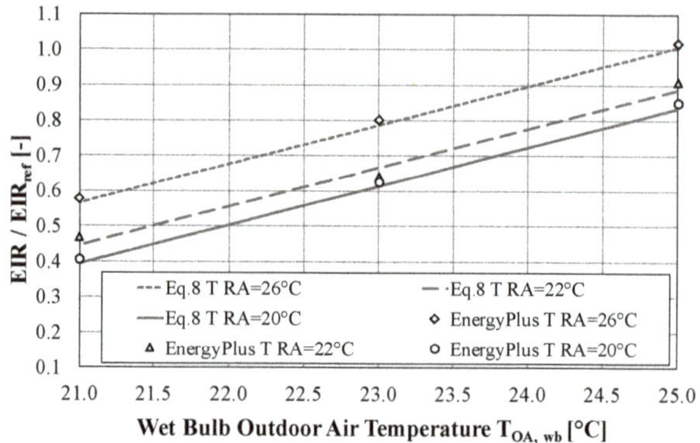

Figure 8. *EIR* in cooling mode: comparison between EnergyPlus simulations and Equation (8) with Table 6 coefficients.

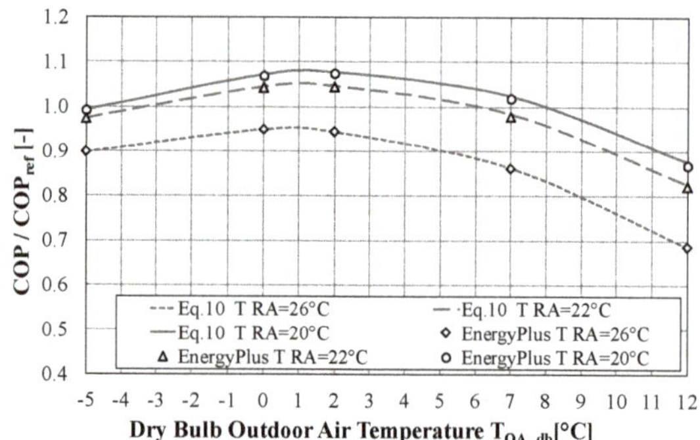

Figure 9. *COP* in heating mode: comparison between EnergyPlus simulations and Equation (10) with Table 6 coefficients.

6. Conclusions

The aim of this work was to provide a series of insights to Energy Plus users when simulations are carried out taking into account the operating temperature effects on performance of heat pumps, chillers and even heat recovery heat pumps in ventilation circuits. The starting point was to refer to the equipment related to a recent near zero energy building at the Authors' University. In particular, the final goal was to properly model the dependence of the heat pumps performance on the temperature, both load and source side and eventually on the partial load operating conditions. The actual installed water-to-water and air-to-air heat pumps have been considered and the equation fit model has been implemented with a series of modifications for adapting it to the typical data available from the manufacturer.

Coefficients needed in the equation fit models have been determined by means of an optimum search and, to validate the approach, a simplified building model equipped with the selected heat pumps and chiller has been created. The results from the simulations confirmed the expected results in terms of heating and cooling equipment performance at given nearly constant working temperatures even if small differences (within 7%) resulted from simulation trends and equation fit model input data. The relative error slightly increases (within 15%) if the partial load operating conditions are considered ($PLF = 0.67, 0.33$).

Author Contributions: The research and actions associated to writing the present paper are equally distributed among the three Authors. All authors have read and agreed to the published version of the manuscript.

Funding: This research received no external funding.

Conflicts of Interest: The authors declare no conflict of interest.

Appendix A

Table A1. Benchmark simulations for the water-to-water heat pump (cooling case).

Operating Conditions				Energy Plus Simulations			Simplified Curve Fit Method		
$T_{L,out}$ [°C]	$T_{S,in}$ [°C]	$T_{L,in}$ [°C]	$T_{S,out}$ [°C]	Electric Power [W]	Cooling Capacity [W]	EER [-]	EER_{ref} [-] (Table 1) Interpolated	T_{ref} [K]	EER [-] Equation (13) with coeff. Table 3
5	20	9.40	24.50	4500	25,699	5.71	5.55	283.15	5.72
5	25	9.18	29.43	5340	24,396	4.57	4.69	283.15	4.58
5	30	9.18	34.61	6544	24,396	3.73	3.98	283.15	3.73
5	35	9.18	39.82	7942	24,395	3.07	3.27	283.15	3.08
5	40	9.18	45.06	9582	24,394	2.55	2.67	283.15	2.55
5	45	9.18	50.35	11,535	24,392	2.11	2.13	283.15	2.12
7	20	11.31	24.38	4240	25,148	5.93	5.80	283.15	5.94
7	25	11.31	29.53	5244	25,148	4.80	4.68	283.15	4.80
7	30	11.31	34.70	6387	25,148	3.94	4.21	283.15	3.94
7	35	11.31	39.89	7701	25,148	3.27	3.51	283.15	3.27
7	40	11.31	45.12	9232	25,166	2.73	2.85	283.15	2.73
7	45	11.31	50.39	11,025	25,162	2.28	2.26	283.15	2.29
12	20	16.48	24.49	4016	26,105	6.50	6.34	283.15	6.51
12	25	16.48	29.62	4904	26,105	5.32	5.42	283.15	5.33
12	30	16.48	34.77	5898	26,105	4.43	4.65	283.15	4.43
12	35	16.48	39.93	7019	26,105	3.72	3.92	283.15	3.72
12	40	16.48	45.12	8293	26,105	3.15	3.18	283.15	3.15
12	45	16.48	50.34	9754	26,105	2.68	2.53	283.15	2.68
15	20	19.50	24.47	3845	26,193	6.81	6.75	283.15	6.82
15	25	19.50	29.60	4664	26,193	5.62	5.79	283.15	5.62
15	30	19.50	34.73	5574	26,193	4.70	4.98	283.15	4.71
15	35	20.07	40.02	6713	26,994	4.02	4.17	283.15	4.03
15	40	20.07	44.95	7505	25,743	3.43	3.45	283.15	3.43
15	45	20.07	50.01	8529	25,085	2.94	2.78	283.15	2.95
18	20	22.49	24.44	3674	26,118	7.11	7.04	283.15	7.12
18	25	22.49	29.55	4430	26,118	5.90	6.07	283.15	5.90
18	30	22.49	34.68	5265	26,118	4.96	5.27	283.15	4.97
18	35	22.49	39.81	6191	26,118	4.22	4.44	283.15	4.22
18	40	22.50	44.98	7244	26,196	3.62	3.63	283.15	3.62
18	45	22.50	50.16	8407	26,196	3.12	2.95	283.15	3.12

Table A2. Benchmark simulations for the water-to-water heat pump (heating case).

Operating Conditions				Energy Plus Simulations			Simplified Curve Fit Method		
$T_{L,out}$ [°C]	$T_{S,in}$ [°C]	$T_{L,in}$ [°C]	$T_{S,out}$ [°C]	Electric Power [W]	Heating Capacity [W]	COP [-]	COP_{ref} [-] (Table 2) Interpolated	T_{ref} [K]	COP Equation (16) with coeff. Table 3
30.0	6	24.6	1.8	10,000	59,092	5.91	5.95	283.15	5.92
30.0	10	24.6	7.4	9045	59,092	6.53	6.62	283.15	6.64
30.0	15	24.6	12.4	8006	59,092	7.38	7.42	283.15	7.49
30.0	18	24.6	13.6	7452	59,092	7.93	7.63	283.15	7.99
30.0	20	24.6	17.3	7108	59,092	8.31	8.28	283.15	8.31
34.6	6	28.7	1.5	12,260	65,200	5.32	5.13	283.15	5.24
34.6	8	28.7	5.2	11,642	65,200	5.60	5.48	283.15	5.57
34.6	12	28.7	9.2	10,525	65,200	6.19	6.04	283.15	6.20
35.0	15	29.0	12.3	9453	62,534	6.62	6.48	283.15	6.61
35.0	20	29.0	17.2	8383	62,534	7.46	7.21	283.15	7.36
45.0	5	38.5	0.9	16,101	63,962	3.97	3.88	283.15	3.92
45.0	6	38.5	1.9	15,649	63,962	4.09	3.98	283.15	4.06
45.0	10	38.5	7.4	14,015	63,962	4.56	4.52	283.15	4.59
45.0	18	38.5	15.3	11,398	63,962	5.61	5.42	283.15	5.61
45.0	20	38.5	17.3	10,848	63,962	5.90	5.68	283.15	5.86
50.0	8	42.8	6.1	13,108	50,297	3.84	3.70	283.15	3.93
49.6	13	46.1	8.3	11,169	43,586	3.90	3.91	283.15	3.90
50.0	10	44.9	10.9	11,911	51,803	4.35	4.16	283.15	4.35
50.0	15	45.1	12.8	11,982	54,673	4.56	4.34	283.15	4.56
50.0	18	45.1	15.8	11,151.4	54,673.3	4.90	4.66	283.15	4.90
50.0	20	45.1	17.7	10,665.7	54,673.3	5.13	4.88	283.15	5.13

Table A3. Benchmark simulations for the air-to-air heat pump (cooling case).

Cooling Mode										
Operating Conditions					Energy Plus Simulations			Simplified Curve Fit Method		
T_{OAdb} [°C]	T_{OAwb} [°C]	T_{RAdb} [°C]	T_{OAwb} [°C]	T_{RAdb} [°C]	Cooling Capacity [W]	Power [W]	EER_S [-]	Cooling Capacity [W]	Power [W]	EER_S [-]
28	21	20	21.04	20.1	22,862.6	3538.5	6.5	23,241.9	3522.0	6.6
28	21	22	21.04	22.2	22,862.6	4062.9	5.6	23,241.9	3996.2	5.8
28	21	26	21.04	25.9	22,862.6	5024.3	4.6	23,241.9	5064.6	4.6
32	23	20	19.84	19.8	33,887.8	8078.1	4.2	33,792.4	7980.2	4.2
32	23	22	20.39	20.4	33,887.8	8264.8	4.1	33,792.4	8669.6	3.9
32	23	26	26.14	26.1	33,887.8	10,314.1	3.3	33,792.4	10,223.0	3.3
40	25	20	25.07	20.1	41,594.1	13,449.5	3.1	41,594.1	13,342.1	3.1
40	25	22	25.07	22.2	40,063.4	13,835.6	2.9	41,594.1	14,190.6	2.9
40	25	26	25.07	25.9	40,063.4	15,510.1	2.6	41,594.1	16,102.7	2.6

Table A4. Benchmark simulations for the air-to-air heat pump (heating case).

Heating Mode									
Operating Conditions				Energy Plus Simulations			Simplified Curve Fit Method		
T_{OAdb} [°C]	T_{RAdb} [°C]	T_{OAdb} [°C]	T_{RAdb} [°C]	Heating Capacity [W]	Power [W]	COP_S [-]	Heating Capacity [W]	Power [W]	COP_S [-]
−5	20	−5.0	20.01	45,426	10,007	4.54	49,292	10,873	4.5
−5	22	−5.0	21.99	45,426	9816	4.63	50,538	10,949	4.6
−5	26	−5.0	26.04	45,426	9071	5.01	50,814	10,147	5.0
0	20	0.0	20.03	46,609	11,077	4.21	47,425	11,298	4.2
0	22	0.0	21.99	46,603	10,787	4.32	48,407	11,241	4.3
0	26	0.0	25.98	46,606	9838	4.74	48,152	10,164	4.7
2	20	2.0	20.01	43,530	10,393	4.19	45,651	10,928	4.2
2	22	2.0	21.94	43,547	10,100	4.31	46,526	10,819	4.3
2	26	2.0	26.03	43,439	9111	4.77	46,059	9661	4.8
7	20	7.0	20.10	35,889	8124	4.42	38,645	8793	4.4
7	22	7.0	22.01	35,885	7804	4.60	39,254	8582	4.6
7	26	7.0	26.30	35,885	6878	5.22	38,258	7332	5.2
12	20	12.0	20.08	28,062	5421	5.18	27,966	5443	5.1
12	22	12.0	21.99	28,206	5144	5.48	28,311	5199	5.4
12	26	12.0	25.93	26,784	4085	6.56	26,784	4085	6.6

References

1. Official Website of the European Union—Energy Performance of Buildings. Available online: https://ec.europa.eu/energy/en/topics/energy-efficiency/energy-performance-of-buildings (accessed on 10 February 2020).
2. Silenzi, F.; Priarone, A.; Fossa, M. Hourly simulations of a hospital building for assessing the thermal demand and the best retrofit strategies for consumption reduction. *Therm. Sci. Eng. Prog.* **2018**, *6*, 388–397. [CrossRef]
3. Linzi, Z.; Joseph, L. Environmental and economic evaluations of building energy retrofits: Case study of a commercial building. *Build. Environ.* **2018**, *145*, 14–23.
4. Liang, J.; Qiu, Y.; James, T.; Ruddell, B.L.; Dalrymple, M.; Earl, S.; Castelazo, A. Do energy retrofits work? Evidence from commercial and residential buildings in Phoenix. *J. Environ. Econ. Manag.* **2018**, *92*, 726–743. [CrossRef]
5. Casquero-Modrego, N.; Goñi-Modrego, M. Energy retrofit of an existing affordable building envelope in Spain, case study. *Sustain. Cities Soc.* **2019**, *44*, 395–405. [CrossRef]
6. Kumar, G.K.; Saboor, S.; Kumar, V.; Kim, K.; Babu, A.T.P. Experimental and theoretical studies of various solar control window glasses for the reduction of cooling and heating loads in buildings across different climatic regions. *Energy Build.* **2018**, *173*, 326–336.
7. Chahine, K.; Murr, R.; Ramadan, M.; El Hage, H.; Khaled, M. Use of parabolic troughs in HVAC applications–Design calculations and analysis. *Case Stud. Therm. Eng.* **2018**, *12*, 285–291. [CrossRef]
8. Soltani, M.; Kashkooli, F.M.; Dehghani-Sanij, A.R.; Kazemi, A.R.; Bordbar, N.; Farshchi, M.J.; Elmi, M.; Gharali, K.; Dusseault, M.B. A comprehensive study of geothermal heating and cooling systems. *Sustain. Cities Soc.* **2019**, *44*, 793–818. [CrossRef]
9. Sebarchievici, C.; Sarbu, I. Performance of an experimental ground-coupled heat pump system for heating, cooling and domestic hot-water operation. *Renew. Energy* **2015**, *76*, 148–159. [CrossRef]
10. Lee, J.; Kim, T.; Leigh, S. Thermal performance analysis of a ground-coupled heat pump integrated with building foundation in summer. *Energy Build.* **2013**, *59*, 37–43. [CrossRef]
11. Qi, D.; Pu, L.; Ma, Z.; Xia, L.; Li, Y. Effects of ground heat exchangers with different connection configurations on the heating performance of GSHP systems. *Geothermics* **2019**, *80*, 20–30. [CrossRef]
12. Lee, S.H.; Jeon, Y.; Chung, H.J.; Cho, W.; Kim, Y. Simulation-based optimization of heating and cooling seasonal performances of an air-to-air heat pump considering operating and design parameters using genetic algorithm. *Appl. Therm. Eng.* **2018**, *144*, 362–370. [CrossRef]
13. Torregrosa-Jaime, B.; Martínez, P.J.; González, B.; Payá-Ballester, G. Modelling of a variable refrigerant flow system in EnergyPlus for building energy simulation in an Open Building Information modelling environment. *Energies* **2019**, *12*, 22. [CrossRef]

14. Fisher, D.E.; Rees, S.J.; Padhmanabhan, S.K.; Murugappan, A. Implementation and validation of ground-source heat pump system models in an integrated building and system simulation environment. *HVAC R Res.* **2006**, *12* (Suppl. 1), 693–710. [CrossRef]

15. Jin, H. Parameter Estimation Based Models of Water Source Heat Pumps. Ph.D. Thesis, Oklahoma State University, Stillwater, OK, USA, 2002.

16. Tang, C. Modeling Packaged Heat Pumps in a Quasi-Steady State Energy Simulation Program. Master's Thesis, Oklahoma State University, Stillwater, OK, USA, 2005.

17. Murugappan, A. Implementing Ground Source Heat Pump and Ground Loop Heat Exchanger Models in the Energyplus Simulation Environment. Master's Thesis, Oklahoma State University, Stillwater, OK, USA, 2002.

18. EnergyPlus Development Team. *EnergyPlus Version 9.1 Engineering Reference: The Reference to EnergyPlus Calculations*; EnergyPlus Development Team, US Department of Energy: Washington, DC, USA, 2019.

19. Fossa, M.; Minchio, F. The effect of borefield geometry and ground thermal load profile on hourly thermal response of geothermal heat pump systems. *Energy* **2013**, *51*, 323–329. [CrossRef]

20. *Zephir CPAN-XHE3: Technical Bullettin*; Clivet: Belluno, Italy, 2016.

Article

Performance and Exergy Transfer Analysis of Heat Exchangers with Graphene Nanofluids in Seawater Source Marine Heat Pump System

Zhe Wang [1,2], Fenghui Han [1,*], Yulong Ji [1] and Wenhua Li [1]

[1]   Marine Engineering College, Dalian Maritime University, Dalian 116026, China
[2]   International Center on Energy Sustainable Ships and Ports, Dalian 116026, China
*   Correspondence: fh.han@dlmu.edu.cn; Tel.: +86-0411-8472-9038

Received: 23 February 2020; Accepted: 31 March 2020; Published: 7 April 2020

Abstract: A marine seawater source heat pump is based on the relatively stable temperature of seawater, and uses it as the system's cold and heat source to provide the ship with the necessary cold and heat energy. This technology is one of the important solutions to reduce ship energy consumption. Therefore, in this paper, the heat exchanger in the CO_2 heat pump system with graphene nano-fluid refrigerant is experimentally studied, and the influence of related factors on its heat transfer enhancement performance is analyzed. First, the paper describes the transformation of the heat pump system experimental bench, the preparation of six different mass concentrations (0~1 wt.%) of graphene nanofluid and its thermophysical properties. Secondly, this paper defines graphene nanofluids as beneficiary fluids, the heat exchanger gains cold fluid heat exergy increase, and the consumption of hot fluid heat is heat exergy decrease. Based on the heat transfer efficiency and exergy efficiency of the heat exchanger, an exergy transfer model was established for a seawater source of tube heat exchanger. Finally, the article carried out a test of enhanced heat transfer of heat exchangers with different concentrations of graphene nanofluid refrigerants under simulated seawater constant temperature conditions and analyzed the test results using energy and an exergy transfer model. The results show that the enhanced heat transfer effect brought by the low concentration (0~0.1 wt.%) of graphene nanofluid is greater than the effect of its viscosity on the performance and has a good exergy transfer effectiveness. When the concentration of graphene nanofluid is too high, the resistance caused by the increase in viscosity will exceed the enhanced heat transfer gain brought by the nanofluid, which results in a significant decrease in the exergy transfer effectiveness.

Keywords: exergy transfer performance; nanofluids; heat exchanger; marine seawater source; heat pump; graphene nanoparticles

1. Introduction

Shipping has always been the most economical mode of transportation for bulk commodities in the world. The global maritime trade relying on ports operates about 80% of the total global trade and transportation [1]. While the shipping industry has effectively promoted the prosperity of global trade and the development of the world economy, it has also brought a lot of environmental pollution. Ship emissions have become a major source of pollution in global port cities and sea areas. According to data from the international maritime organization, if no effective energy-saving and emission reduction measures are taken at present, the CO_2 emissions of marine vessels will account for 18% of the global total by 2050 [2]. Most of the ship's energy consumption and emissions are energy consumption for its cargo storage, personnel production and life. Traditional ship air conditioners generally use exhaust gas or oil-fired boilers to generate saturated steam above 150 °C for heat exchangers. Then, the fan or steam pipeline sends the heat to each cabin, to achieve the purpose of heating and domestic hot water.

However, due to the general slowdown of the ship, the steam generated by the exhaust gas boiler cannot meet the needs of the whole ship, and an oil-fired boiler has to be used to assist [3], which will lead to an increase in the operating cost of the whole ship. Data show that the demand for refrigeration air conditioning and hot water for cruise ships accounts for more than 45% of the entire ship's electricity consumption [4]. These electric power needs to consume a large amount of fuel, which will generate huge additional energy consumption in addition to power navigation, especially when the ship's power system is not working after the port is docked, and its energy consumption and emissions will be more serious without the main engine energy recovery [5]. In summary, the quality of the air in the ship, as well as the supply of cold and heat, will not only greatly affect the health and efficiency of the crew and passengers, but also bring safety risks to the operation of the cargo and the main and auxiliary equipment of the ship's operation.

Based on the existing problems in current ship equipment facing the severe world maritime ship environmental protection, active global energy conservation and emission reduction policy guidance, ship energy-saving heat pump technology and research came into being. Marine seawater source heat pumps take advantage of the relatively stable temperature of the ocean [6] and use the compressor in the system to extract low-grade energy from seawater at the expense of a small amount of electrical energy. This energy is used as a cold/heat source of the heat pump system, and its temperature is increased or decreased before being transmitted to the user, to provide the required cold/heat or domestic hot water on board. Its energy efficiency ratio of winter heating can generally reach 3~6, while the summer cooling energy efficiency ratio can reach 2~4 [7]. Traditional heating ventilation air conditioning (HVAC) requires a lot of auxiliary equipment to complete the above functions, and the marine heat pump system of seawater source only needs to switch between the seawater side condenser and the evaporator according to the user's needs to complete the cooling and heating function adjustment. Therefore, the heat pump technology will be an important solution to reduce the ship's electrical energy consumption, navigation costs and emissions. Additionally, heat pump technology has an important application value in many cases. Fossa [8] studied the application of borehole heat exchangers in ground-coupled heat pumps. The basic thermal response factor is recursively calculated, and a direct method for calculating the long-term effective ground resistance is given. Andrei et al. [9] introduced the use of a seawater source heat pump and analyzed related energy consumption to illustrate the temperature gradient utilization of marine heat pumps and the advantages of higher energy efficiency. Zheng et al. [10] tested the heat transfer performance of a polyethylene spiral coil heat exchanger in a heat pump system with a seawater source. The effects of seawater flow and surface icing on the heat transfer performance of the heat exchanger were studied. Liu et al. [11] gave a heat pump system with seawater source with a capillary tube as a front-end heat exchanger. Studies have found that capillary tubes serve as a heat exchanger between external seawater and internal glycol antifreeze, reducing initial system investment and increasing COP. Ezgi et al. [12] investigated the feasibility of using steam jet heat pump systems in ships. The optimal design parameters were obtained by performing a thermodynamic analysis using an H_2O-LiBr absorption heat pump as the HVAC system. Yun et al. [13] proposed an automatic cascade heat pump system to overcome the adverse effect of environmental temperature on the efficiency of the CO_2 heat pump. The automatic cascade heat pump uses two-stage expansion and CO_2-R32 azeotropic refrigerant. Priarone et al. [14] used a curve fitting method to import measured data from a large number of suppliers into the software, and built models of ground source water heat pumps and air heat pumps, and finally estimated the COP of the heat pump under different temperature operating conditions.

How to select the working fluid of the thermal system and evaluate the energy consumption of the heat exchange equipment has always been an important content of engineering practice design and optimization improvement research. Firstly, the working medium circulating in the heat pump system exchanges energy with the outside world through the change of its thermal state, to realize reverse Carnot cycle and other energy transfer from low temperature to high temperature. The performance of its system depends to some extent on the characteristics of the working medium. Secondly, due to environmental protection requirements, various classification societies currently restrict the use of traditional HFC refrigerants [15]. Therefore, the selection of a suitable environmentally friendly working medium is crucial to the performance of the ship's heat pump system and energy saving and emission reduction. Chen et al. [16] studied the distribution of R134a refrigerant in the system under different operating conditions to improve the efficiency of the vapor compression refrigeration unit. The results can be used to optimize the refrigerant charge of the refrigeration unit and its components. Bobbo et al. [17] analyzed different refrigerant fluids as alternatives to the ground source heat pump R410A. The thermodynamic properties, flow rate, pressure, and heat transfer efficiency of R32, R290, and R454b are compared in the system compressor, evaporator, and condenser under given conditions. Paniagua et al. [18] compared the system performance of transcritical CO_2 and R410A heat pumps in zero-energy comprehensive building applications. Combined with experimental data and models, the situation of two hot pumps receiving and generating hot water at the same ambient temperature was analyzed. Aiming at the characteristic that transcritical CO_2 is more sensitive to working conditions, an optimization method was developed to improve its performance. The above research content reflects an alternative situation of the current thermal system working fluid. The research hotspots are mainly the application of environmentally friendly new working fluid and the impact of the system.

There are many analysis and evaluation methods currently used to measure the energy consumption of heat exchangers and guide their optimal design [19]. They can be divided into the following three categories: first, the analysis and evaluation based on the first law of thermodynamics [20]; second, the evaluation combining the conservation of energy and the second law of thermodynamics [21]; finally, the above methods are improved and derived, including thermoeconomics [22], composition method [23], the theory of fire accumulation dissipation [24], etc. Sun et al. [25] studied a shell-and-tube heat exchanger with inclined three-lobed baffles, numerically simulated its flow and heat transfer characteristics, and analyzed the structure using heat transfer coefficients and pressure drops. Shen et al. [26] considered the limitations of traditional methods in the evaluation of mechanical energy dissipation, introduced the theory of entropy production, and intuitively studied the mechanical energy dissipation of variable flow and blade tip clearance through numerical simulation of axial flow pumps. Preißinger et al. [27] studied the design of ORC systems, the selection of working fluids and the matching of heat sources and heat sink temperatures from the perspective of thermal economics, and developed ORC thermoeconomic models based on complexity parameters and structural dimensions. In summary, most of the evaluation of heat exchange equipment is for equipment improvement and performance improvement, but most of them only involve fluid flow, heat transfer performance and relatively little analysis of the overall energy transfer and loss of the equipment.

In addition to the above studies, the heat exchanger is also an important part of the system. Especially for seawater source heat pumps, the heat exchanger is a key component that directly affects the use of low-grade energy and conversion performance [28]. To prevent seawater from corroding the heat exchange equipment and damaging the ship's power and related systems, the condenser or evaporator of a marine heat pump system with seawater source, in general, does not directly exchange heat with seawater. The refrigerant vapor in the heat pump evaporator first transfers the cooling capacity to a certain kind of refrigerant, and then carries the energy transfer to the seawater through the intermediate heat exchanger. In this process, how to ensure the efficient transfer of energy in various heat exchangers has been a problem in the industry [29]. Therefore, it is a current research hotspot to use nanoparticles as a refrigerant to enhance the heat transfer method to actively increase the flow heat transfer rate of the heat exchanger and increase the energy efficiency coefficient of the heat pump system [30]. For the design and optimization of heat exchangers with different structures and new heat exchange fluids in a complex heat exchange system, traditional models and correlations cannot fully accurately reflect the effects of heat transfer and pressure drop on system performance. Therefore, specific problems need to be analyzed and studied in detail for working conditions, structures and new fluid conditions. Aprea et al. [31] investigated the effect of nanofluids as refrigerant auxiliary fluids on heat transfer enhancement in vapor compression refrigeration systems. The results show that adding 10% copper nanoparticles to the water/glycol mixture can enhance the heat transfer of the medium by about 30%. Kristiawan et al. [32] studied two passively enhanced heat transfer technologies combining a microchannel structure and a nanofluid by measuring Nusselt number, friction coefficient and heat transfer performance. It has been proven that, compared with water flowing in a square microchannel, using nanofluids with a volume fraction of 0.01% can increase the heat transfer efficiency of the microchannel. Ramirez et al. [33] numerically studied the forced convection process of five nanofluids in straight tubes of different structures under constant temperature and constant heat flux conditions. The results show that as the volume fraction of Al_2O_3 nanoparticles increases, the average Nusselt number increases. Qi et al. [34] used a two-step method to prepare stable titanium dioxide-water nanofluids and tested their heat transfer and flow characteristics in triangular and circular tube heat exchanger systems. The fitting formula of the Nusselt number and drag coefficient of nanofluid in the triangular tube is given, and the comprehensive performance of nanofluid in the triangular tube is studied. In recent years, research on related nanometers has been very rapid. Due to its better thermophysical properties, the heat transfer performance of the fluid can be increased. Among many nanofluids, graphene is a leader with higher thermal conductivity and electrical conductivity, and smaller viscosity [35–37]. Therefore, it is very promising to study graphene nanofluids as refrigerants to increase the performance of traditional thermal systems.

Given the above problems, this article draws on the advantages of various parties in the literature to conduct experimental research and performance analysis on a heat exchanger with graphene nano-fluid refrigerant in an environmentally friendly working fluid transcritical CO_2 heat pump system. The temperature and pressure coupled exergy transfer analysis method is used to specifically evaluate the comprehensive performance of the heat exchanger to guide subsequent optimization and practical design.

2. Research Methods and Experiments

2.1. Heat Pump Heat Exchanger Test-Bed

The purpose of this article is to study the role of graphene nanofluids as refrigerants in improving the performance of heat exchangers in seawater source heat pump systems. We retrofit the experimental system of the previous marine transcritical CO_2 heat pump system with a seawater source. The original heat pump system is a 1.5 HP transcritical CO_2 water source heat pump system test bench for heating energy efficiency of about four, which can supply 80–200 L of hot water at a temperature of 55~60 °C [38,39]. The modified seawater source circulation system is shown in Figure 1. Based on the nano-fluid refrigerant-enhanced heat transfer enhanced seawater source heat pump heat exchanger, a testbed for improving performance, other transcritical CO_2 heat pump systems, data measurement acquisition and electrical control systems are constant.

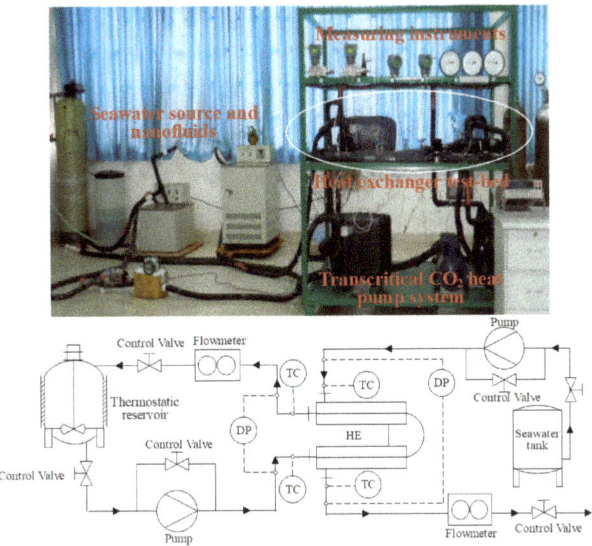

Figure 1. Heat pump system with seawater source and heat exchanger testbed.

The modified nano-fluid heat exchanger experimental process is as follows. Firstly, a nano-fluid refrigerant is charged in a constant temperature, stirred tank and a constant temperature is set to simulate the heat transfer conditions of the heat pump evaporator. Subsequently, the refrigerant in the constant temperature tank is transported by the circulation pump to the tube side of the coaxial sleeve heat exchanger at a constant flow rate to exchange heat with the seawater. After measuring its temperature difference, pressure drop and flowrate, it returns to the constant temperature tank to complete the heat exchange cycle on the heat exchanger tube side. Finally, the seawater is transported from the storage tank to the shell side of the coaxial sleeve heat exchanger by the open pump to exchange heat with the nanofluids on the tube side. After measuring the temperature difference, pressure drop, and flowrate, the seawater is discharged from the open cycle, and the heat exchange on the shell side of the heat exchanger ends. The structure and heat exchange area of the related test heat exchangers is shown in Table 1. The shell-side material of the heat exchanger uses stainless steel to prevent seawater corrosion and pressure. The nickel-copper pipe is used on the tube side of the heat exchanger to enhance heat exchange. The error analysis of relevant parameters about this experiment system was shown in Table 2. All the measurement parameters were recorded by a computer with the data logger system. Error analysis of measurements is calculated by Equation (1), where the x_1, x_2, etc. and w_1, w_2, etc. are the independent variables of a measuring result and the uncertainties, respectively.

$$Er = \left(\left(\frac{\partial r}{x_1} w_1 \right)^2 + \left(\frac{\partial r}{x_2} w_2 \right)^2 + \ldots + \left(\frac{\partial r}{x_n} w_n \right)^2 \right)^{0.5} \tag{1}$$

Table 1. Dimensions of the heat exchangers.

Items	Coaxial Tube Heat Exchanger
Structure tube ID/OD	Double-pipe 16 single row 9.5 mm/16 mm
Total heat exchange area	1.9 m^2
Material ID/OD	Stainless steel/Nickel cupronickel

Table 2. Measuring the operating parameters of the instrument.

Items	Units	Measuring Range	Accuracy
Temperature	°C	−35.0–200.0	±0.2
Pressure	bar	0–100	±1.0%
Pressure drop	bar	0–10	±1.0%
Sea water flow rate	m^3/h	0–5	±0.5%
Refrigerant mass flow rate	kg/s	0–0.5	±0.5%
Pump power	kW	0–10	±0.5%

2.2. Properties of Graphene Nanofluid

As shown in Figure 2, a scanning electron microscope (SEM) was used to observe the micrograph of the purchased graphene nanofluid stock solution 5 wt.%. The SEM images of the GnP were provided by the supplier, who gives the GnP images at 15 μm, 10 μm, and 5 μm magnified sizes. It can be seen from the SEM images that, due to the special two-dimensional structure of the graphene sheet, the graphene nanosheets are distributed in a wrinkled shape in the base fluid solution. Figure 3 is a process for preparing nanofluids with different concentrations and testing their thermophysical properties. Firstly, a 5 wt.% graphene nanofluid stock solution purchased from a company in the USA and its glycol-based solution are shown. After being diluted by a certain mass ratio, there are five kinds of nanofluids to be tested (0.01~1 wt.%) with different mass concentrations. Secondly, we added a surfactant to each of the nanofluids to be measured, and put them into an ultrasonic vibration water bath to diffuse the nanoparticles in the solution, adjusting the duration so that the nanosheets completely diffuse into the base liquid. Finally, the dispersion stability of the solution was tested by zeta potential analysis, and then the viscosity and thermal conductivity of the fluids were measured by a rotational viscometer and a thermal constant analyzer.

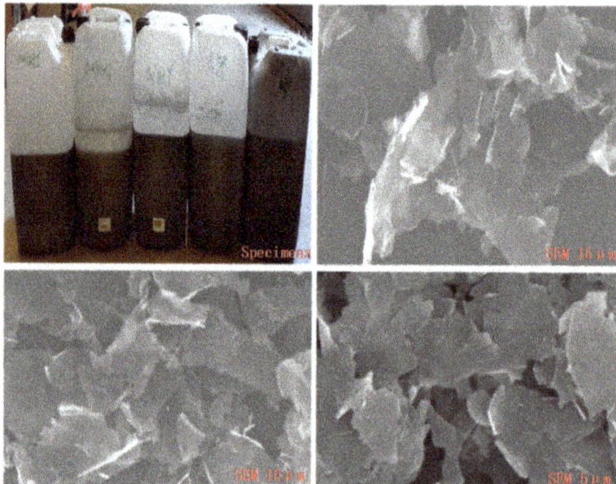

Figure 2. Nanofluids samples views of the experiment and SEM image of GnP.

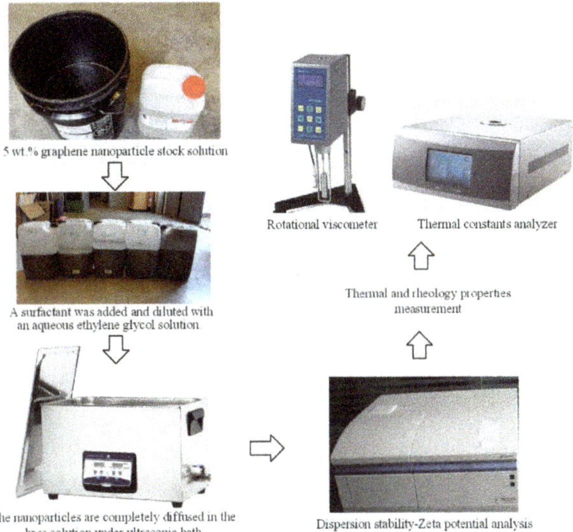

Figure 3. Graphene nanofluid preparation and the physical property measurement process.

In summary, the thermal properties of the five graphene nanofluids are shown in Table 3. It can be seen that as the concentration of nanoparticles increases, the thermal conductivity and viscosity coefficient of the fluid to be measured are also increasing. When the mass concentration reaches 1.0, the viscosity of the graphene nanofluid will increase rapidly, and the value measured by the rotational viscometer exceeds three times that of the base fluid. Additionally, their density and specific heat capacity can be given according to the following empirical formulas. For detailed processes and analysis, refer to references [40,41].

$$\rho_{nf} = (1 - \phi)\rho_{bf}(T) + \phi\rho_p \tag{2}$$

$$\rho_{nf}c_{p,nf} = (1 - \phi) \cdot \rho_{bf}(T)c_{p,bf}(T) + \phi\rho_p c_{p,np} \tag{3}$$

where the ρ and cp are density and specific heat of the GnP nanofluids, the ϕ indicates the volume fraction of the nanoparticles, and subscripts *nf*, *bf* and *np* indicate nanofluid, base fluid and nanoparticles, respectively.

Table 3. Thermal properties of nanofluids compared to base fluids.

Item					
Mass fraction	0.01	0.05	0.1	0.5	1.0
Volume fraction	0.0048%	0.024%	0.048%	0.24%	0.48%
k_{nf}/k_{bf}	1.017	1.019	1.021	1.102	1.211
μ_{nf}/μ_{bf}	1.001	1.034	1.058	1.488	3.0

3. Exergy Transfer Model and Data Reduction

The traditional evaluation of the energy efficiency of heat exchangers and the use of nanofluids to enhance heat transfer analysis are limited to changes in related parameters such as Nusselt number, Reynolds number, heat transfer rate, pumping and convective heat transfer coefficients [42,43]. To dig deep into the energy transfer, utilization and loss caused by the use of nanofluids in heat exchangers, this paper uses the idea of heat exchanger effectiveness to build an exergy transfer model using exergy analysis and exergy transfer theory. It is expected to reveal the intrinsic energy transfer, utilization and loss of nanofluids in the enhanced heat transfer process in heat exchangers from the perspective of exergy transfer.

The exergy loss in the heat exchanger includes two parts caused by the limited heat transfer temperature difference and the fluid viscous flow resistance. Therefore, it is assumed that the flow and heat transfer processes in the heat exchanger are in a stable state, the axial heat conduction and external heat dissipation are ignored, the nanofluid working medium is stable in physical properties, and the working temperature is above the ambient temperature. The differential expression of the specific exergy change of the working medium in the heat exchanger in this paper is given as:

$$de = c_p(1 - \frac{T_\Theta}{T})dT + [v - (T - T_\Theta)(\frac{\partial v}{\partial T})_P]dP \tag{4}$$

where, c_p is specific heat at constant pressure, v is the specific volume of the working medium, T_Θ is the environment temperature. For incompressible fluids, v is considered to be constant in normal physical properties and $(\frac{\partial v}{\partial T})_P = 0$ is brought into Equation (4) to obtain:

$$de = c_p(1 - \frac{T_\Theta}{T})dT + vdP \tag{5}$$

By integrating the Equation (5), it can obtain:

$$\begin{aligned} \Delta E &= \int_{P_i,T_i}^{P_o,T_o} mc_p(1 - \frac{T_\Theta}{T})dT + \int_{P_i,T_i}^{P_o,T_o} mvdP \\ &= mc_p(T_o - T_i - T_\Theta \ln \frac{T_o}{T_i}) - mv\Delta P \end{aligned} \tag{6}$$

$$\Delta P = f\frac{L\rho u^2}{2d} \tag{7}$$

where P is pressure, T is temperature, m is mass flow of medium, f is the friction factor, u is the velocity of the medium, d is the equivalent diameter, ρ is the density of the working medium, L is the length of the heat exchanger, and the subscripts i and o are the inlet and outlet of the heat exchanger. After obtaining the above expression, the exergy transfer effectiveness ε_e of the heat exchanger is defined as the actual exergy change of the target heat exchange medium compared to its maximum possible exergy change. For the heat exchanger of this study, the target medium is cold fluid, which benefits from the increase of the cold fluid (graphene nanofluid refrigerant) heat exergy and its consumption is the reduction of the hot fluid (seawater) heat exergy. The maximum possible exergy change of the medium is the maximum exergy change in the ideal state of the countercurrent, i.e., the case of the pressure exergy loss of the countercurrent is 0, after that:

$$\varepsilon_E = \frac{m_c c_{pc}(T_\Theta - T_{ci} - T_\Theta \ln \frac{T_\Theta}{T_{ci}}) - m_c v_c \Delta P}{m_h c_{ph}(T_{hi} - T_{ci} - T_\Theta \ln \frac{T_{hi}}{T_{ci}})} \tag{8}$$

Define a series of dimensionless numbers:

$$D = \frac{m_c c_{p,c}}{m_h c_{p,h}} = \frac{T_{h,i} - T_{h,o}}{T_\Theta - T_{c,i}}, \ \varphi = \frac{T_\Theta - T_{c,i}}{T_{h,i} - T_{c,i}}, \ \gamma = \frac{T_{c,i}}{T_{h,i}}, \ \gamma_0 = \frac{T_\Theta}{T_{h,i}} \tag{9}$$

then

$$T_\Theta = T_{c,i} + \varphi(T_{h,i} - T_{c,i}), \ T_{h,o} = T_{h,i} - \varphi D(T_{h,i} - T_{c,i}) \tag{10}$$

putting into Equation (8) to simplify, it can get:

$$\varepsilon_E = \frac{\varphi(1 - \gamma) - \gamma_0 \ln[1 + \gamma(1 - \gamma)\gamma] - \frac{\gamma_0}{c_{p,c}\rho_c T_\Theta}\Delta P}{1 - \gamma + \gamma_0 \ln \gamma} \tag{11}$$

where D is the heat capacity ratio of the cold and hot fluid, and φ is the heat transfer effectiveness of the heat exchanger. Relevant formulas used in other analyses are shown in Table 4.

Table 4. Mathematical formulas used in the correlation analysis of this study.

NO.	Items	Note
1	$d = \frac{4s}{L}$	Equivalent diameter:
2	$Q = \frac{(q_h + q_c)}{2}$	Heat flux
3	$\text{LMTD} = \frac{(T_{h,i} + T_{c,o}) - (T_{h,o} + T_{c,i})}{\ln[(T_{h,i} - T_{c,o})/(T_{h,o} - T_{c,i})]}$	Logarithmic mean temperature difference
4	$h = \frac{Q}{A\,\text{LMTD}}$	Convective heat transfer coefficient
5	$\varphi = \frac{Ah\,\text{LMTD}}{(mc_p)_{min}(T_{h,i} - T_{c,i})}$	Thermal efficiency of heat exchanger
6	$P_w = \frac{m\Delta P}{\rho}$	Pump power
7	$\text{Nu} = \frac{hd}{k}$	Nusselt number
8	$\text{Re} = \frac{\rho u d}{\mu}$	Reynolds number
9	$\text{Pr} = \frac{\mu c_p}{k}$	Prandtl number
10	$\text{Pe} = \frac{\rho c_p L u}{k}$	Peclet number
11	$\text{NTU} = \frac{KA}{(mc_p)_{min}}$	Number of Transfer Units

4. Results and Discussion

4.1. Performance Analysis of Heat Exchanger

To clarify the influence of nanofluids on the enhancement of convective heat transfer, the changing trend of the ratio of heat conduction resistance to convective heat transfer resistance at the bottom of laminar fluid flow was explored. Six groups (1 wt.%, 0.5 wt.%, 0.1 wt.%, 0.05 wt.%, 0.01 wt.%, 0 wt.%) of graphene nanofluids with different mass fractions were used for the following experimental study, and their base fluid was 50 wt.%:50 wt.% ethylene glycol aqueous solution. As shown in Figure 4, its influence on the heat transfer performance of coaxial sleeve heat exchanger and the change of Nusselt number with Reynolds number. The results show that the increase of the Reynolds number increases the disturbance of the fluid and destroys the flow boundary layer. The more intense the convective heat transfer of the fluid in the tube, the stronger its heat transfer performance. Additionally, as the concentration of graphene nanoparticles increases, the heat transfer performance also improves. It can be explained that the thermal conductivity of the nanoparticles is significantly higher than that of the base fluid, so as the concentration of the nanofluid increases, the thermal conductivity of the working fluid will be greater than that of the base fluid. In addition, graphene nanoparticles are subjected to Brownian forces in nanofluids to perform irregular Brownian diffusion and thermal diffusion movement [40–42]. It makes micro-convection between the nanoparticles and the base fluid, enhances the energy transfer between the nanoparticles and the base fluid, sharpens the destruction of the boundary layer, enhances disturbance, and enhances heat transfer. However, with the increase in the concentration of graphene nanoparticles, the viscosity of the nanofluid continues to increase. If a high Nusselt number is to be maintained, a large amount of pump work is consumed [42]. In the perspective of energy conservation, it is not economical to increase the mass fraction of graphene nanoparticles to obtain a small thermal conductivity gain. From the change trend of the ratio of the thermal resistance to the convective heat transfer resistance at the bottom of the fluid laminar flow, the heat transfer enhancement rate from 0.01% to 0.1% is greater than the heat transfer enhancement rate from 0.1% to 1%. These results can increase the heat transfer gain by about 7% or more.

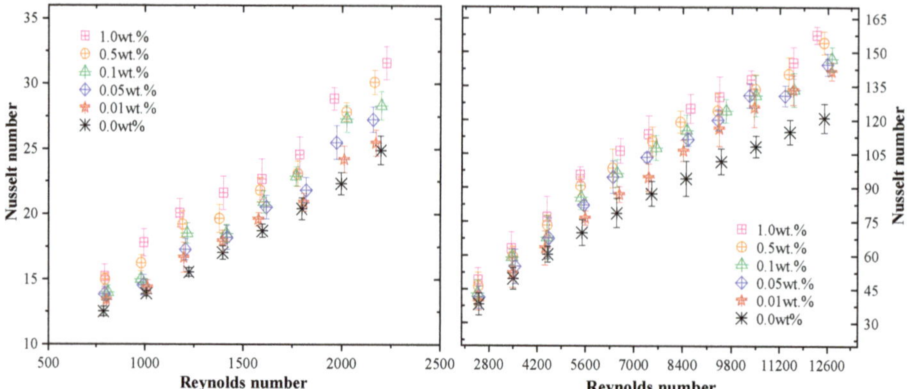

Figure 4. Effects of nanofluids mass fraction on Nusselt number in a laminar and turbulent flow.

The effect of graphene nanofluids with different mass fractions on the friction resistance coefficient of coaxial sleeve heat exchangers is shown in Figure 5. The root mean square error (RMSE) measured during the experiment is less than ±7.4%, and the relevant trend is well reflected by the error bars and relevant test points. It is clear that the friction factor increases as the mass fraction of graphene nanoparticles increase at similar Reynolds numbers. It is because the friction factor is positively related to the viscosity of the fluid, and the viscosity of the nanofluid increases with the increase of the nanoparticle mass fraction, so the friction factor increases. In addition, in laminar and turbulent flows, the friction factor decreases as the Reynolds number increases. This is due to the negative correlation between the friction factor and the square of the flow velocity. Therefore, in addition to changing the viscosity of the fluid, increasing the nanoparticle concentration will not cause other properties to affect the friction factor.

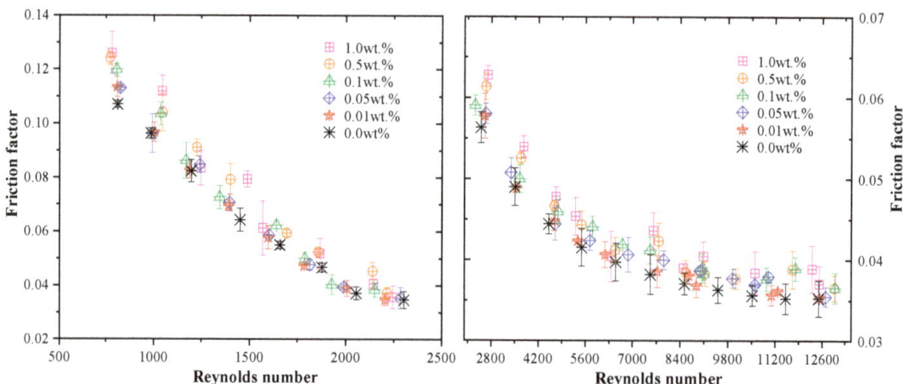

Figure 5. Effects of nanofluids mass fraction on friction factor in laminar and turbulent flow.

The above analysis only demonstrated the positive effects of increasing flow rate and graphene nanoparticle concentration on characterizing enhanced heat transfer. However, the corresponding friction factor will also increase, and the energy efficiency evaluation of the heat exchanger is simply the result of strengthening the trade-off between the enhanced heat transfer and drag coefficient [40,41]. To further quantitatively analyze the effects of different graphene nanofluid concentrations and flow rates on the performance of heat exchangers, Figure 6 shows the relationship between the pump work and thermal efficiency of the heat exchanger. The results show that, under the experimental conditions, the heat transfer effectiveness of the six measured nanofluids in the heat exchanger gradually increases with the increase of the pump work. The increase in heat exchange performance of the heat exchanger is often accompanied by a pressure drop or pump power consumption, which is in line with the first law of thermodynamics. Secondly, due to the increase in the heat transfer effectiveness of the heat exchanger caused by different concentrations of graphene nanofluids, the increase in unit pump work is greater than the increase in heat exchange capacity. The heat transfer effectiveness of graphene nanofluids at the concentration of 0.01~0.1 wt.% was greater than 0.5~1 wt.%. This indicates that graphene nanofluids with a concentration above 0.1 wt.% in the heat exchanger can improve the heat transfer by means of a greater thermal conductivity of the fluid, but also cause greater frictional resistance and consume more pump work.

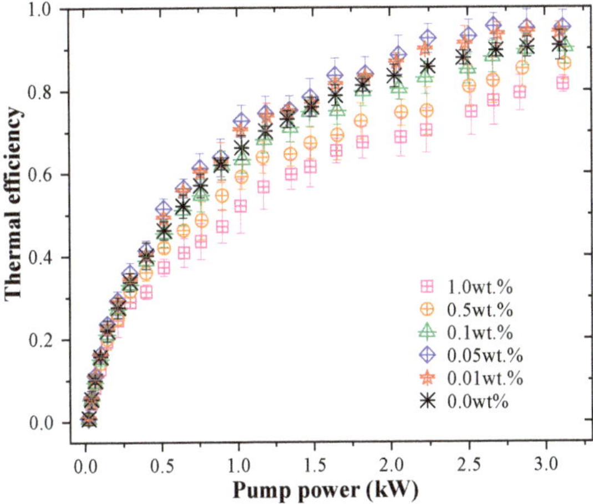

Figure 6. Pump power versus thermal efficiency of heat exchanger.

Figure 7 shows the relationship between Peclet number on the nanofluid side of the heat exchanger and the convective heat transfer coefficient of the fluid. The heat transfer Peclet number is the product of the Reynolds number and Prandtl number. Its physical meaning is the ratio of convective heat transfer to fluid heat transfer under forced motion. It takes into account both the momentum and the heat diffusivity, i.e., it also measures the density, specific heat, thermal conductivity, and flowrate. The effect of this dimensionless number on the convection heat transfer coefficient of the heat exchanger while increasing the thermal conductivity of the fluid is now given. The results show that as the Peclet number increases, the convective heat transfer coefficients of all measured fluids gradually increase. At the concentration of 0.01~0.1 wt.%, the convective heat transfer coefficient of graphene nanofluids was higher than that of base fluid (0 wt.%). It is because the heat transfer improvement effect brought by the addition of nanoparticles is greater than the increase in its thermal resistance. The convective heat transfer coefficient at the concentration of 0.5~1 wt.% was lower than that of the base fluid, and the thermal resistance of the heat exchanger on the nanofluid side increases sharply. In this interval, the increase of the nanofluid concentration causes the convective heat transfer coefficient to decrease.

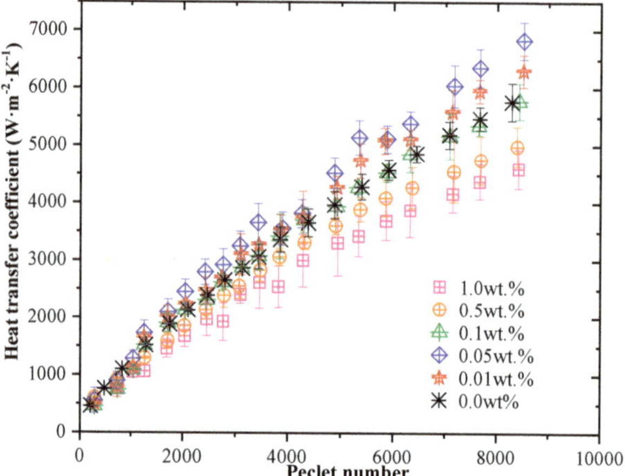

Figure 7. Convective heat transfer coefficient versus the Peclet number.

4.2. Energy Transfer Efficiency Analysis

The exergy loss in the heat exchanger includes two parts caused by the limited heat transfer temperature difference and the fluid viscous flow resistance. Therefore, the exergy transfer function is used to analyze the influence of the two factors on the energy transfer of the heat exchanger, i.e., the enhanced heat transfer caused by the thermal conductivity of the nanofluid and the loss of pump power caused by its viscosity increase. Figure 8 shows the changing trend of the influence of different concentrations of graphene nanofluids on the exergy transfer effectiveness of the heat exchanger under different NTU conditions. The NTU defines the ratio of the total heat conductance of the fluid (i.e., the inverse of the heat transfer heat resistance of the heat exchanger) to the heat capacity of the fluid, which is an indicator of the comprehensive technical and economic efficiency of the reaction heat exchanger. Since the experimentally tested heat exchanger has a certain volume, it only changes the flow rate and heat capacity of the hot and cold fluids. Therefore, under countercurrent conditions, as the NTU increases, the exergy transfer effectiveness of the graphene nanofluid side in the heat exchanger gradually increases. It can be expected that the final value will approach one, which is the case of infinite flow velocity. Besides, when at a certain NTU value, the heat exchanger has a better exergy transfer effectiveness when the nanoparticle concentration is in the range of 0.01~0.1 wt.% compared to the range of 0.5~1 wt.%. It is because the viscosity of the heat transfer medium and the pump work lost gradually increase as the concentration increases, and the effect of related energy dissipation of graphene nanofluids at concentrations above 0.1 wt.% exceeds the benefits of enhanced heat transfer.

Figure 9 shows the effect of the concentration of graphene nanofluid on the heat transfer effectiveness of heat exchangers under different conditions of heat capacity ratio of cold and hot fluid. In this study, the graphene nanofluid was defined as the benefit fluid, and the increase in heat exergy of the cold fluid was obtained, while the decrease in heat exergy of the hot fluid was consumed. The exergy transfer effectiveness of graphene nanofluids with different concentrations showed different variation trends, and the heat exchangers with concentrations of 0.01~0.1 wt.% had better performance than those with concentrations of 0.5~1 wt.%. It is because the exergy loss includes two parts caused by the limited heat transfer temperature difference and the fluid viscous flow resistance. The enhanced heat transfer effect brought by the lower concentration of graphene nanofluid is greater than the effect of its fluid viscosity on performance. When the concentration of the graphene nanofluid is too high, the resistance caused by the increase in viscosity will exceed the enhanced heat transfer gain brought by the nanofluid, leading to a significant decrease in the exergy transfer effectiveness, which meets the definition of Equation (11). Besides, the exergy transfer effectiveness of graphene nanofluids in heat exchangers decreases with increasing D and eventually approaches a constant value. It is because for a non-phase change heat exchanger, the smaller the change in the heat capacity ratio of the cold and hot fluid in the heat exchanger is equivalent to the smaller the temperature difference between the heat exchanger and the outside, the greater the corresponding effectiveness. Therefore, when designing a single-phase heat exchanger without phase change, it is important that the hot and cold fluids have an approximate heat capacity ratio so that a high exergy transfer effectiveness can be achieved.

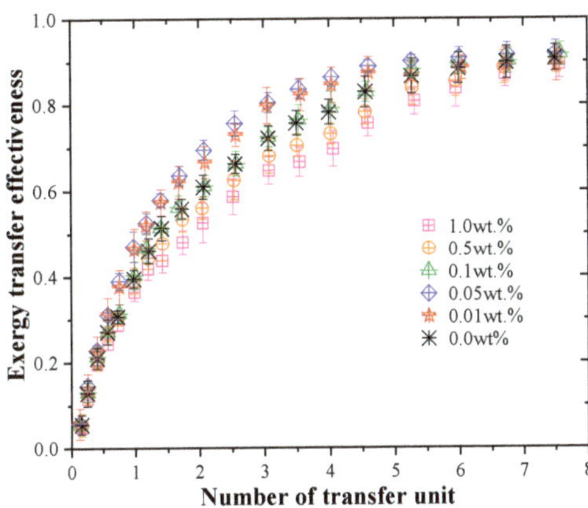

Figure 8. Effects of nanofluids mass fraction on exergy transfer effectiveness at different NTU.

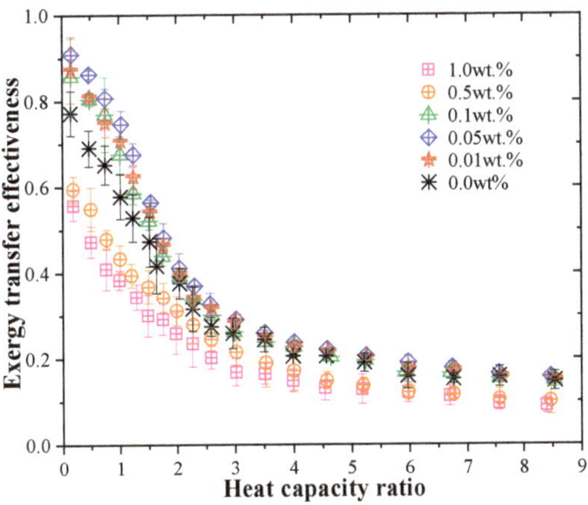

Figure 9. Effects of nanofluids mass fraction on exergy transfer effectiveness at different heat capacity ratios.

5. Conclusions

This paper focuses on the heat pump technology for the energy saving and emission reduction of maritime ships and carries out the experimental test and analysis based on the first and second laws of thermodynamics for a heat exchanger with a graphene nanofluids refrigerant in a transcritical CO_2 heat pump system. Firstly, the effect of the enhanced heat transfer and fluid resistance factor in the heat exchanger using different concentrations of graphene nanofluids was studied. Secondly, based on the heat transfer effectiveness and exergy efficiency of the heat exchanger, a heat exchanger exergy transfer model was theoretically established. Finally, exergy transfer was evaluated for various working conditions of the graphene nanofluid heat exchanger with different concentrations. The main conclusions are as follows:

(1) Based on the comparison of equal Re numbers under laminar and turbulent conditions, the thermophysical properties of GnP nanofluids are the main factors affecting the heat transfer performance and fluid resistance factor of heat exchangers. Under the same Re number conditions, both the Nu number and the heat transfer factor increase to varying degrees with the increase of nanoparticle concentration.

(2) According to the further analysis of the first law of thermodynamics, it can be obtained that the proper increase of pump work and Pe number can significantly increase the heat transfer efficiency and convective heat transfer coefficient of the nanofluid. The heat transfer efficiency of the graphene nanofluids at unit pump work was greater than 0.5~1 wt.%, and the convective heat transfer coefficient of the graphene nanofluids was higher than that of the base fluid at 0.01~0.1 wt.%.

(3) The exergy transfer effectiveness of the heat exchanger in this study defines the graphene nanofluid as the benefit fluid, gains an increase in the exergy of the cold fluid heat, and consumes a decrease in the exergy of the hot fluid heat. The graphene nanofluids with concentrations of 0.01~0.1 wt.% had better exergy transfer effectiveness than those with concentrations of 0.5~1 wt.% in the heat exchanger. The enhanced heat transfer effect of the lower concentration of graphene nanofluid is greater than the effect of its viscosity on performance. When the concentration of graphene nanofluid is too high, the resistance caused by the increase in viscosity will exceed the enhanced heat transfer gain brought by the nanofluid, leading to a significant decrease in the exergy transfer effectiveness.

Author Contributions: Z.W. and F.H. conducted the experiments and wrote the paper. Z.W. analyzed the results. Y.J. gave some suggestions and carried on project administration/funding acquisition. W.L. commented on the paper. All authors have read and agreed to the published version of the manuscript.

Funding: This work was funded by the Fundamental Research Funds for the Central Universities (No. 3132020, 3132019331), the National Natural Science Foundation of China (No. 51906026). Their support is gratefully acknowledged.

Conflicts of Interest: The authors declare no conflict of interest.

Nomenclature

A	area of surface-heat transfer (m^2)
c_p	heat capacity (J·kg^{-1}·K^{-1})
d	hydraulic diameter (m)
D	heat capacity ratio
e	exergy (W)
ΔE	specific exergy (W)
f	friction factor
h	heat transfer coefficient (W·m^{-2}·K^{-1})
L	heat transfer length (mm)
k	thermal conductivity (W·m^{-1}·K^{-1})
m	mass flow rate (kg·s^{-1})
Nu	Nusselt number
ΔP	pressure drop (kPa)
Pe	Peclet number
Pr	Prandtl number
q	heat (W)
Q	heat flux (W)
Re	Reynolds number
s	cross-sectional area (m^2)
T	temperature (K)
U	overall heat transfer coefficient (W·m^{-2}·K^{-1})
v	specific volume (m^3·kg^{-1})

P_w	pump power (kW)
μ	dynamic viscosity (Pa·s)
u	velocity (m·s^{-1})
Greek symbols	
ε_E	exergy transfer effectiveness
ρ	density (kg·m^{-3})
ϕ	volume concentration
φ	thermal efficiency
Subscripts	
bf	base fluid
exp	experimental
c, h	cold or hot side
nf	nanofluid
i, o	inlet or outlet
Θ	environment
Abbreviations	
EGW	ethylene glycol-water
GnP	graphene nanoplatelets
CHTC	convective heat-transfer coefficient
ID/OD	inner tube/outer tube
HVAC	heating ventilation air conditioning
TC	thermocouple
DP	differential pressure
HE	heat exchanger
RMSE	root-mean-square error

References

1. Han, F.; Wang, Z.; Ji, Y.; Li, W. Energy analysis and multi-objective optimization of waste heat and cold energy recovery process in LNG-fueled vessels based on a triple organic Rankine cycle. *Energy Convers. Manag.* **2019**, *195*, 561–572. [CrossRef]
2. Trivyza, N.L.; Rentizelas, A.; Theotokatos, G. Impact of carbon pricing on the cruise ship energy systems optimal configuration. *Energy* **2019**, *175*, 952–966. [CrossRef]
3. Benamara, H.; Hoffmann, J.; Youssef, F. Maritime Transport: The Sustainability Imperative. In *Sustainable Shipping*; Springer: Cham, Switzerland, 2019; pp. 1–31.
4. Goldsworthy, B.; Goldsworthy, L. Assigning machinery power values for estimating ship exhaust emissions: Comparison of auxiliary power schemes. *Sci. Total Environ.* **2019**, *657*, 963–977. [CrossRef] [PubMed]
5. Valencia Ochoa, G.; Acevedo Peñaloza, C.; Duarte Forero, J. Thermoeconomic optimization with PSO Algorithm of waste heat recovery systems based on organic rankine cycle system for a natural gas engine. *Energies* **2019**, *12*, 4165. [CrossRef]
6. Zhao, Z.; Zhang, Y.; Mi, H.; Zhou, Y.; Zhang, Y. Experimental research of a water-source heat pump water heater system. *Energies* **2018**, *11*, 1205. [CrossRef]
7. Valarezo, A.S.; Sun, X.Y.; Ge, T.S.; Dai, Y.J.; Wang, R.Z. Experimental investigation on performance of a novel composite desiccant coated heat exchanger in summer and winter seasons. *Energy* **2019**, *166*, 506–518. [CrossRef]
8. Fossa, M. The temperature penalty approach to the design of borehole heat exchangers for heat pump applications. *Energy Build.* **2011**, *43*, 1473–1479. [CrossRef]
9. Andrei, P. *Study Regarding Marine Heat Pump*; Analele Universitatii Maritime Constanta: Constanta, Romania, 2012; Volume 13, pp. 143–146.
10. Zheng, W.; Zhang, H.; You, S.; Ye, T. Numerical and experimental investigation of a helical coil heat exchanger for seawater-source heat pump in cold region. *Int. J. Heat Mass Transf.* **2016**, *96*, 1–10. [CrossRef]
11. Liu, L.; Wang, M.; Chen, Y. A practical research on capillaries used as a front-end heat exchanger of seawater-source heat pump. *Energy* **2019**, *171*, 170–179. [CrossRef]

12. Ezgi, C.; Girgin, I. Design and thermodynamic analysis of a steam ejector refrigeration/heat pump system for naval surface ship applications. *Entropy* **2015**, *17*, 8152–8173. [CrossRef]

13. Yun, S.K. Study on the performance characteristics of a new CO_2 auto-cascade heat pump system. *J. Korean Soc. Mar. Eng.* **2017**, *41*, 191–196. [CrossRef]

14. Priarone, A.; Silenzi, F.; Fossa, M. Modelling Heat Pumps with Variable EER and COP in EnergyPlus: A Case Study Applied to Ground Source and Heat Recovery Heat Pump Systems. *Energies* **2020**, *13*, 794. [CrossRef]

15. Hafner, I.A.; Gabrielii, C.H.; Widell, K. *Refrigeration Units in Marine Vessels: Alternatives to HCFCs and High GWP HFCs*; Nordic Council of Ministers: Copenhagen, Denmark, 2019; pp. 1–90.

16. Chen, X.; Li, Z.; Zhao, Y.; Jiang, H.; Liang, K.; Chen, J. Modelling of refrigerant distribution in an oil-free refrigeration system using R134a. *Energies* **2019**, *12*, 4792. [CrossRef]

17. Bobbo, S.; Fedele, L.; Curcio, M.; Bet, A.; De Carli, M.; Emmi, G.; Poletto, F.; Tarabotti, A.; Mendrinos, D.; Mezzasalma, G.; et al. Energetic and exergetic analysis of low global warming potential refrigerants as substitutes for R410A in ground source heat pumps. *Energies* **2019**, *12*, 3538. [CrossRef]

18. López Paniagua, I.; Jiménez Álvaro, Á.; Rodríguez Martín, J.; González Fernández, C.; Nieto Carlier, R. Comparison of transcritical CO_2 and conventional refrigerant heat pump water heaters for domestic applications. *Energies* **2019**, *12*, 479. [CrossRef]

19. Wang, Z.; Li, Y. Layer pattern thermal design and optimization for multistream plate-fin heat exchangers—A review. *Renew. Sustain. Energy Rev.* **2016**, *53*, 500–514. [CrossRef]

20. Zhao, Z.; Zhou, Y.; Ma, X.; Chen, X.; Li, S.; Yang, S. Effect of different zigzag channel shapes of PCHEs on heat transfer performance of supercritical LNG. *Energies* **2019**, *12*, 2085. [CrossRef]

21. Pons, M. Exergy Analysis and process optimization with variable environment temperature. *Energies* **2019**, *12*, 4655. [CrossRef]

22. Lillo, G.; Mastrullo, R.; Mauro, A.W.; Trinchieri, R.; Viscito, L. Thermo-economic analysis of a hybrid ejector refrigerating system based on a low grade heat source. *Energies* **2020**, *13*, 562. [CrossRef]

23. Bejan, A.; Lorente, S. Constructal theory of generation of configuration in nature and engineering. *J. Appl. Phys.* **2006**, *100*, 041301. [CrossRef]

24. Alic, F. Entransy dissipation analysis and new irreversibility dimension ratio of nanofluid flow through adaptive heating elements. *Energies* **2020**, *13*, 114. [CrossRef]

25. Sun, Y.; Wang, X.; Long, R.; Yuan, F.; Yang, K. Numerical Investigation and Optimization on Shell Side Performance of a Shell and Tube Heat Exchanger with Inclined Trefoil-Hole Baffles. *Energies* **2019**, *12*, 4138. [CrossRef]

26. Shen, S.; Qian, Z.; Ji, B. Numerical Analysis of Mechanical Energy Dissipation for an Axial-Flow Pump Based on Entropy Generation Theory. *Energies* **2019**, *12*, 4162. [CrossRef]

27. Preißinger, M.; Brüggemann, D. Thermoeconomic evaluation of modular organic Rankine cycles for waste heat recovery over a broad range of heat source temperatures and capacities. *Energies* **2017**, *10*, 269. [CrossRef]

28. Wu, Z.; You, S.; Zhang, H.; Fan, M. Mathematical Modeling and Performance Analysis of Seawater Heat Exchanger in Closed-Loop Seawater-Source Heat Pump System. *J. Energy Eng.* **2019**, *4*, 04019012. [CrossRef]

29. Alam, T.; Kim, M.H. A comprehensive review on single phase heat transfer enhancement techniques in heat exchanger applications. *Renew. Sustain. Energy Rev.* **2018**, *81*, 813–839. [CrossRef]

30. Sun, X.H.; Yan, H.; Massoudi, M.; Chen, Z.H.; Wu, W.T. Numerical simulation of nanofluid suspensions in a geothermal heat exchanger. *Energies* **2018**, *11*, 919. [CrossRef]

31. Aprea, C.; Greco, A.; Maiorino, A.; Masselli, C. Enhancing the heat transfer in an active barocaloric cooling system using ethylene-glycol based nanofluids as secondary medium. *Energies* **2019**, *12*, 2902. [CrossRef]

32. Kristiawan, B.; Wijayanta, A.T.; Enoki, K.; Miyazaki, T.; Aziz, M. Heat Transfer Enhancement of TiO2/water nanofluids flowing inside a square minichannel with a microfin structure: A numerical investigation. *Energies* **2019**, *12*, 3041. [CrossRef]

33. Ramirez-Tijerina, R.; Rivera-Solorio, C.; Singh, J.; Nigam, K.D.P. Numerical study of heat transfer enhancement for laminar nanofluids flow. *Appl. Sci.* **2018**, *8*, 2661. [CrossRef]

34. Qi, C.; Liu, M.; Wang, G.; Pan, Y.; Liang, L. Experimental research on stabilities, thermophysical properties and heat transfer enhancement of nanofluids in heat exchanger systems. *Chin. J. Chem. Eng.* **2018**, *26*, 2420–2430. [CrossRef]

35. Ghozatloo, A.; Rashidi, A.; Shariaty-Niassar, M. Convective heat transfer enhancement of graphene nanofluids in shell and tube heat exchanger. *Exp. Therm. Fluid Sci.* **2014**, *53*, 136–141. [CrossRef]

36. Sadeghinezhad, E.; Mehrali, M.; Saidur, R.; Mehrali, M.; Latibari, S.T.; Akhiani, A.R.; Metselaar, H.S.C. A comprehensive review on graphene nanofluids: Recent research, development and applications. *Energy Convers. Manag.* **2016**, *111*, 466–487. [CrossRef]

37. Arshad, A.; Jabbal, M.; Yan, Y.; Reay, D. A review on graphene based nanofluids: Preparation, characterization and applications. *J. Mol. Liq.* **2019**, *279*, 444–484. [CrossRef]

38. Wang, Z.; Gong, Y.; Wu, X.H.; Lu, Y. Thermodynamic analysis and experimental research of transcritical CO_2 cycle with internal heat exchanger and dual expansion. *Int. J. Air Cond. Refrig.* **2013**, *21*, 1350005. [CrossRef]

39. Wang, Z.; Han, F.; Sundén, B. Parametric evaluation and performance comparison of a modified CO_2 transcritical refrigeration cycle in air-conditioning applications. *Chem. Eng. Res. Des.* **2018**, *131*, 617–625. [CrossRef]

40. Wu, Z.; Wang, L.; Sunden, B. Pressure drop and convective heat transfer of water and nanofluids in a double-pipe helical heat exchanger. *Appl. Therm. Eng.* **2013**, *60*, 266–274. [CrossRef]

41. Wang, Z.; Wu, Z.; Han, F.; Wadsö, L.; Sundén, B. Experimental comparative evaluation of a graphene nanofluid coolant in miniature plate heat exchanger. *Int. J. Therm. Sci.* **2018**, *130*, 148–156. [CrossRef]

42. Tariq, R.; Zhan, C.; Ahmed Sheikh, N.; Zhao, X. Thermal performance enhancement of a cross-flow-type Maisotsenko heat and mass exchanger using various nanofluids. *Energies* **2018**, *11*, 2656. [CrossRef]

43. Wang, Z.; Li, Y. Irreversibility analysis for optimization design of plate fin heat exchangers using a multi-objective cuckoo search algorithm. *Energy Convers. Manag.* **2015**, *101*, 126–135. [CrossRef]

 *energies*

Article

Operation of a Tube GAHE in Northeastern Poland in Spring and Summer—A Comparison of Real-World Data with Mathematically Modeled Data

Aldona Skotnicka-Siepsiak

Faculty of Geoengineering, University of Warmia and Mazury in Olsztyn, 10-724 Olsztyn, Heweliusza, Poland; aldona.skotnicka-siepsiak@uwm.edu.pl

Received: 27 February 2020; Accepted: 3 April 2020; Published: 7 April 2020

Abstract: The article analyzes a ground-to-air heat exchanger (GAHE) for a mechanical ventilation system in a building. The heat exchanger's performance was evaluated in northeastern Poland between May and August of 2016, 2017, and 2018. In spring and summer, the GAHE can be theoretically used to precool air for HVAC systems. The aim of the study was to compare the real-world performance of GAHE with its theoretical performance determined based on the distribution of ground temperature and the temperature at the GAHE outlet modeled in compliance with Standard PN-EN 16798-5 1:2017-07. The modeled values differed considerably from real-world data in May and June, but the model demonstrated satisfactory data fit in July and August. In all years, the modeled average monthly air temperature at the GAHE outlet was 8.3 °C below real-world values in May, but the above difference was only 1.1 °C in August. The developed mathematical model is simple and easy to use, and it can be deployed already in the preliminary design stage. It does not require expensive software or expert skills. However, this study revealed that the model has several limitations. The observed discrepancies should be taken into account when modeling the performance of a GAHE.

Keywords: ground-to-air heat exchangers; GAHE; experimental results; preheating and precooling for HVAC; energy saving for HVAC; models for calculating the thermal efficiency of ground-to-air heat exchangers

---

1. Introduction

A ground-to-air heat exchanger (GAHE) is a relatively simple technology which can be incorporated with conventional heating, ventilation, and air-conditioning (HVAC) installations to preheat and precool air [1]. Despite its simplicity, a GAHE can effectively lower the demand for indoor heat and minimize the environmental impact of heating installations [2].

In Poland, GAHEs are increasingly often used to preheat air, both in small ventilation systems in single-family homes and in large installations where the diameter of ground tubes exceeds 900 mm [3]. The performance of a GAHE in the winter of 2016 was analyzed by [4]. The experimentally measured values were compared with theoretical computational models based on Standard PN-EN 15241:2011 [5,6]. A comparison of outdoor air temperature measured experimentally at the GAHE outlet with the modeled data revealed that the average monthly heating load was 23% higher in the theoretical model. Even greater discrepancies were observed when typical meteorological year (TMY) data were used in the calculations, where the average heat gain was 34% higher in the model than in the experimental measurements. These differences can be attributed mainly to the fact that the temperatures of outdoor air in the TMY dataset were significantly below the measured temperatures, in particular in winter. This observation is not surprising in the face of global climate change.

A GAHE can also be used to precool air, and such applications have been analyzed in research studies conducted in Iraq [7], India [8], Algeria [9], China [10], and Turkey [11]. However, not only sensible heat transfer—but also latent heat exchange—has to be taken into account when air precooled in a GAHE is fed to HVAC installations [12].

The GAHE concept and design continues to be studied, optimized and developed. The current state of knowledge and the most recent developments have been presented in several review articles [13–15]. Some authors have investigated selected aspects of GAHE design and operation. Misra et al. [16] proposed a solution for minimizing GAHE installation costs and maximizing energy gain. Baglivo et al. [17] aimed to determine the optimal number of tubes in a horizontal GAHE by analyzing air temperature, air flow rate, pipe depth, and soil thermal conductivity in TRNSYS 17 software. GAHEs can be integrated with solar chimneys for passive cooling of buildings [18].

The operation, design, and optimization of GAHEs are often analyzed with the use of numerical simulations [19–21]. Kumar et al. [22] relied on a genetic algorithm to optimize the design of a GAHE. The authors compared a deterministic model with an intelligent design model. The intelligent model, which accounted for variations in the humidity of circulating air, natural thermal stratification of the ground, latent and sensible heat transfer, as well as ground conditions, proved to be a more effective solution. Ahmed et al. [23] used the Fluent software package to develop a model for the Australian climate which revealed that the cooling load of a GAHE can be optimized by increasing the depth at which the tubes are buried in the ground, maximizing the tubes' thermal conductivity, and minimizing tube diameter and the thickness of tube walls. The optimal air flow rate in the analyzed system was 1.5 m/s. Rodrigues et al. [24] relied on the constructal design method to optimize the thermal capacity of a GAHE. Bisoniya [25] described the application of the ε-NTU method for designing an effective GAHE installation. Rouag, Benchabane, and Mehdid [26,27] developed a semi-analytical method to determine the distance from the pipe axis in the ground (soil radius) influenced by heat from the GAHE. Unlike other researchers, the authors assumed that the soil radius affected by heat from the GAHE is not stable and that it is influenced by the duration of operation, soil thermal diffusivity, pipe diameter, and air temperature.

Despite the availability of numerous CFD methods, there is a demand for simple computational models for engineers who do not have access to advanced numerical tools. A simple method for estimating the preheating and precooling requirements of a HVAC installation coupled to a GAHE has been proposed by Chiesa [28]. In the cited study, the model's effectiveness was evaluated with the use of three key performance indicators (KPI): an analysis of the number of operating hours based on a psychrometric chart, the anticipated sensible heat exchange (in winter and summer), and the determination of a virtual coefficient of performance based on a theoretically calculated pressure drop. Peretti et al. [29] discussed the design and environmental impacts of a GAHE in a review article. The authors reviewed the existing models and algorithms for evaluating heat transfer in soil and the thermal behavior of pipes, and they discussed several real-world applications of GAHEs. The article presents formulas for determining the heating and cooling loads of a GAHE, including formulas for calculating the interactions between ground and ambient air, heat transfer equations, and a mathematical model for determining the annual subsurface soil temperature. Peretti applied a mathematical model of ground temperature developed by [30]. Other mathematical models have been proposed by ASHRAE [31]; Tittelein, Achard, and Wurtz [32]; Badache et al. [33]; and Ozgener, Ozgener, and Tester [34]. An empirical model for calculating soil temperature was described by Popiel and Wojtkowiak [6].

The aim of this study was to compare the real-world performance and the theoretical performance of a GAHE based on the results of an experiment and a mathematical model developed in accordance with Standard PN-EN 16798-5 1:2017-07 *"Energy Performance of Buildings—Ventilation for Buildings—Part 5-1: Calculation Methods for Energy Requirements of Ventilation and Air Conditioning Systems (Modules M5-6, M5-8, M6-5, M6-8, M7-5, M7-8)—Method 1: Distribution and Generation"*. The experimental setup for real-world measurements is presented in the next section. The mechanical ventilation system in the presented experiment supported both air precooling and preheating. The mathematical model was developed based on Standard PN-EN 16798-5 1:2017-07 to determine temperature distribution in the ground. The above standard was also applied to calculate air temperature at the GAHE outlet (the temperature of outdoor air was determined based on TMY data and the measurements conducted in the summers of 2016, 2017, and 2018). The modeled data were compared with the results of the experimental measurements of GAHE performance.

2. Materials and Methods

Laboratory analyses were carried out in the Institute of Civil Engineering of the University of Warmia and Mazury in Olsztyn (Poland). The experimental setup involved an AwaduktThermo ground-to-air heat exchanger (GAHE) buried in the ground to a depth of 1.97 m at a point where the tube crossed the wall of a building to 2.27 m by a water drainage tank. The experimental setup is presented graphically in Figure 1. The tubes were buried in the ground (wet sand) with a downward slope in the direction of the water drainage tank located in the proximity of the AwaduktThermo air intake stack. Total tube length was 41 m, and internal tube diameter was 0.2 m.

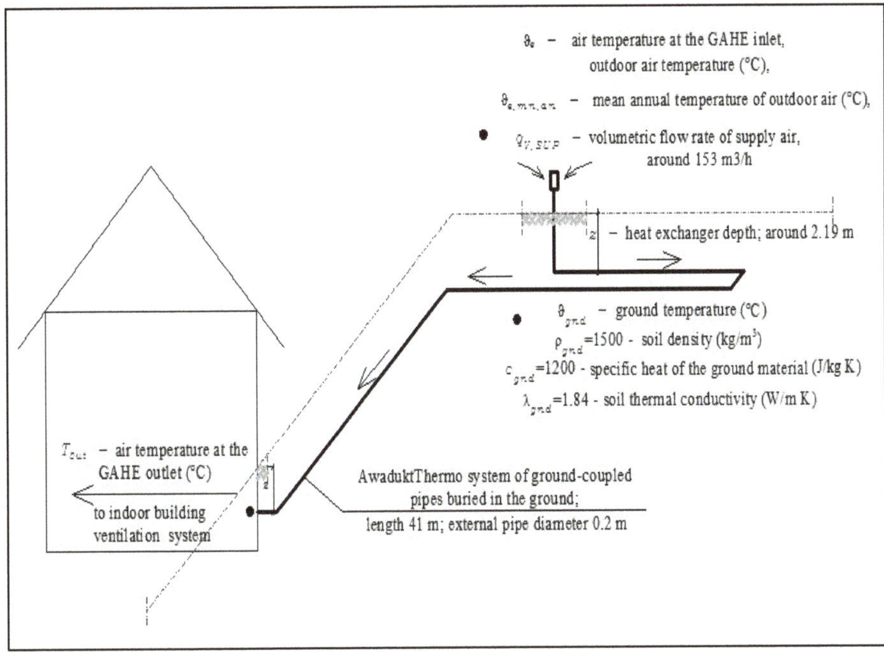

Figure 1. Diagram of the experimental setup.

AwaduktThermo tubes were the main element of the experimental setup. The tubes have antibacterial internal lining, and the base polymer is enhanced with silver particles to reduce microbial contamination in the system. The antibacterial properties of internal tube lining have been tested and certified by the Polish National Institute of Public Health and Institut Fresenius in Taunusstein, Germany. The tubes have solid polypropylene walls for optimal heat transfer [35].

The measurements were conducted with resistance temperature detectors and humidity sensors installed outdoors and in the GAHE at the point where the tubes entered the building. Temperature at the GAHE inlet and outlet was measured with the Siemens QAM2120.040 duct temperature sensor with a measuring range of −50 °C to +80 °C. The resistance of the sensing element changes as a function of temperature, and measuring accuracy is ±0.4 K at 0 °C and ±0.5 K at 20 °C. The device is equipped with an LG-Ni 1000 sensing element with a nominal resistance of 1000 Ω/0 °C, and it has a time constant of less than 20 s during assembly in a pipeline. Humidity was measured with the Siemens QFM2100 duct sensor with a measuring range of 0% to 100%, and measuring accuracy of ±5% at 23 °C and 24 V AC. Air flow meters were installed in the GAHE, and a pyranometer was also used. The pyranometer has a spectral range of 300 nm to 2800 nm, output voltage of 5 mV/W/m^2 to 20 mV/W/m^2, response time of 18 s, and directional error of less than 20 W/m^2. The air flow rate was determined with the Siemens QVM62.1 air velocity sensor with a measuring range of 0 to 10 m/s and measuring accuracy of ±0.2 m/s (+3% of the measured value) at 20 °C, 45% humidity, and atmospheric pressure of 1013 hPa. Sensor data were registered by a Siemens controller in real time and were averaged at hourly intervals.

The measurements were conducted between May and August of 2016, 2017, and 2018. Meteorological data were compared with typical meteorological year (TMY) data, and 2952 measurements were obtained in every analyzed year. The number of measurements conducted at the GAHE outlet was determined by the system's operating time (Figure 2). In 2016, the GAHE operated continuously (2952 h), and the average air flow rate was approximately 163 m^3/h. In the remaining years of the experiment, the GAHE operated intermittently. In May 2017, the system operated at weekly intervals. In June 2017, three operating days were followed by a five-day pause. In July and August 2017, the GAHE operated on working days only, for around 8 h per day (7:00 a.m. to 3:00 p.m.). In 2018, the GAHE operated continuously in May and in the first half of June. Between mid-June and mid-July 2018, the system operated intermittently at 7-day to 14-day intervals. From mid-July to the end of August 2018, the GAHE operated on working days only, for around 8 h per day (7:00 a.m. to 3:00 p.m.). In 2017, total operating time was 1038 h with an average air flow rate of around 150 m^3/h. In 2018, total operating time was 1608 h with an average air flow rate of around 145 m^3/h.

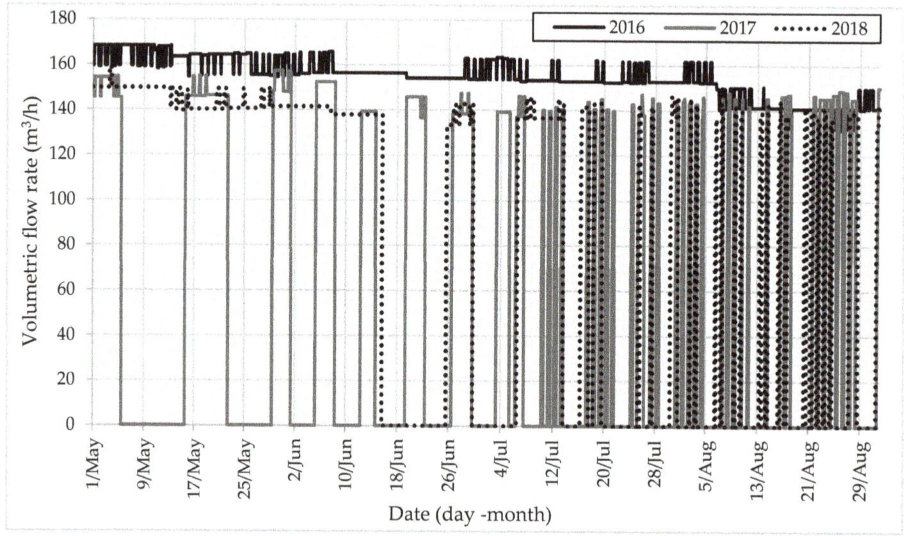

Figure 2. GAHE operating time.

The heating and cooling loads of the GAHE were calculated with the following formula [36]

$$Q = \dot{m} \cdot c_p \cdot (T_{Out} - T_{In}) \tag{1}$$

where:

Q—heat gain from the GAHE (W);

$\dot{m}$—mass flow rate of air (kg/s);

c_p—specific heat of air (J/(kg·K));

T_{In}—air temperature at the GAHE inlet (outdoor air) (K);

T_{Out}—air temperature at the GAHE outlet (K).

A positive result indicates that heat was transferred from the ground-to-air, whereas a negative result indicates that air was cooled in the GAHE. Heating and cooling loads were determined on an hourly basis, and the results were expressed in Wh.

However, Equation (1) requires real-world data that have to be obtained through experimental measurements. Air temperature at the GAHE outlet is most difficult to measure at the design stage. The difference in air temperature between the GAHE inlet and outlet can be determined based on Standard PN-EN 16798-5-1:2017-07

$$\Delta\vartheta_{gnd} = \left(\vartheta_{gnd} - \vartheta_e\right)\left[1 - e^{-\left(\frac{U_{du}\cdot A_s}{q_{V;SUP}\rho_a c_a}\right)}\right] \tag{2}$$

where:

ϑ_e—outdoor air temperature (°C);
A_s—inner surface area of the GAHE (m^2);
$q_{V;SUP}$—volumetric flow rate of supply air (m^3/h);
ρ_a—air density (kg/m^3);
c_a—specific heat of air at constant pressure, ($c_a = 0.000279$ kWh/(kg K)).
Ground temperature ϑ_{gnd} is calculated with the formula

$$\vartheta_{gnd} = \vartheta_{e;mn;an} + \left(\vartheta_{e;max;m} - \vartheta_{e;mn;an}\right)\cdot e^{-\xi}\cdot\cos\left(2\pi\cdot\frac{t_{an}}{8760} - \xi - f_t\right) \tag{3}$$

where:

$\vartheta_{e;mn;an}$—mean annual temperature of outdoor air (°C);
$\vartheta_{e;max;m}$—maximum mean monthly temperature of outdoor air (°C);
t_{an}—hours per year (h).
Coefficient ξ accounts for soil type and the depth of GAHE tubes

$$\xi = z\cdot\sqrt{\frac{\pi\cdot\rho_{gnd}\cdot c_{gnd}}{\lambda_{gnd}\cdot 8760\cdot 3600}} \tag{4}$$

where:

ρ_{gnd}—soil density, (1500 kg/m^3);
c_{gnd}—specific heat of the ground material, (1200 J/kg K);
λ_{gnd}—soil thermal conductivity, (1.88 W/m K);
z—tube depth (m).
The flow time coefficient f_t is given by the formula

$$f_t = \pi\left(\frac{2\cdot t_{an;,in}}{8760} + 1\right) \tag{5}$$

where:

$t_{an;min}$—number of hours in the year with minimal mean monthly temperature of outdoor air (h).
The overall heat transfer coefficient of the GAHE is expressed by the formula

$$U_{du} = \left(\frac{1}{2\pi}\cdot\frac{1}{\lambda_{du}}\cdot\ln\frac{\frac{d_o}{2}}{\frac{d_i}{2}} + \frac{1}{h_i}\right)^{-1} \tag{6}$$

where:

λ_{du}—thermal conductivity of the tube, (0.28 W/m K);
d_i—inner diameter of the tube, (0.200 m);

d_o—outer diameter of the tube, (0.188 m).

The inside surface coefficient h_i is calculated as

$$h_i = \left[4.13 + 0.23 \cdot \frac{\vartheta_m}{100} - 0.0077 \cdot \left(\frac{\vartheta_m}{100} \right)^2 \right] \cdot \frac{v^{0.75}}{d_i^{0.25}}$$

(7)

where:

ϑ_m—average air temperature inside the tube (°C);

v—air velocity inside the tube (m/s).

Ground temperature and the difference in air temperature between the GAHE inlet and outlet in the analyzed period (May to August 2016, 2017, and 2018) were determined with the use of a theoretical mathematical model compliant with Standard PN-EN 16798-5-1:2017-07 and formulas (2) to (7). The calculations were performed for the measured temperatures of outdoor air and the measured air flow rates in each year of the experiment. The theoretical heating and cooling loads of the GAHE were determined with the use of formula (1). The results produced by the theoretical model were compared with the results of the calculations performed with the use of formula (1) based on the experimentally measured values.

Latent heat transfer, namely the energy released during water vapor condensation, was determined with the use of formula (8) at hourly intervals [37]. The results were expressed in Wh.

$$Q_t = \dot{m} \cdot (h_{Out} - h_{In})$$

(8)

where:

Q_t—latent heat transfer, (W);

$\dot{m}$—mass air flow rate (kg/s);

h_{In}—enthalpy of fresh air at the GAHE inlet (J/kg);

h_{Out}—enthalpy of fresh air at the GAHE outlet (J/kg).

Air enthalpy was calculated with the below formula:

$$h = c_d \cdot t + (q + c_v \cdot t) \cdot d$$

(9)

where:

h—air enthalpy (J/kg);

c_d—average specific heat of dry air at constant pressure, (1001 J/(kg K));

t—air temperature (K);

q—vaporization heat of water at 0 °C, (2,500,000 J/kg);

c_v—average specific heat of water vapor at constant pressure, (1840 J/(kg K));

d—specific humidity (kg/kg).

3. Results

Temperatures exceeded TMY values in each year of the study. The average TMY temperature in the analyzed period was 15.1 °C, whereas the average annual temperature during the experiment was determined at 16.6 °C in 2016, 17.0 °C in 2017, and 19.2 °C in 2018 (Figure 3). In the analyzed period, the lowest TMY temperature was −3.2 °C and the highest TMY temperature was 31.0 °C. During the experiment, the corresponding values were determined at 3.4 °C and 32.1 °C in 2016, 0.1 °C and 29.0 °C in 2017, and 4.8 °C and 31.5 °C in 2018. The above trend was also observed in analyses of monthly data.

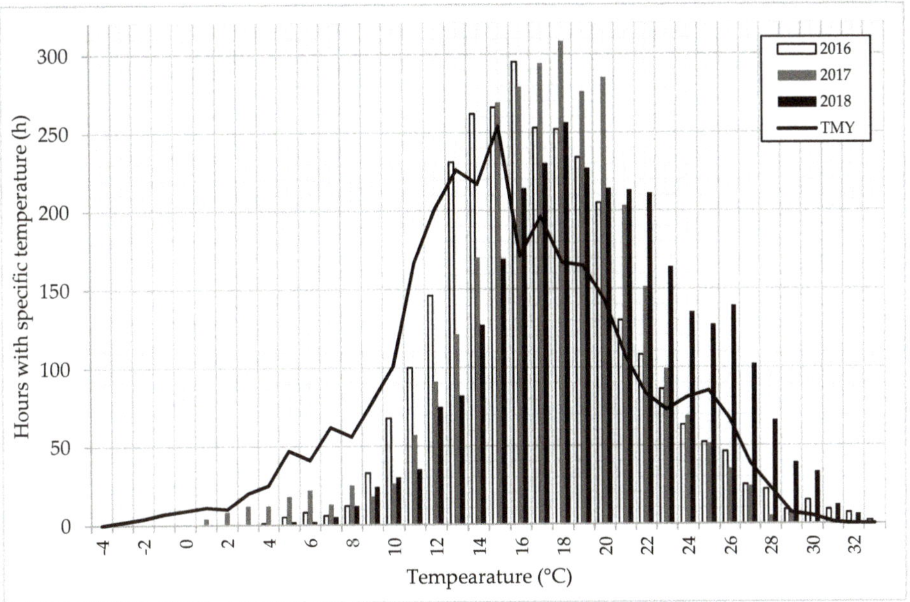

Figure 3. Distribution of outdoor air temperature in the analyzed period.

Total solar irradiance (TSI) in the analyzed period was determined at 515,867 W/m^2 based on TMY data, and the measured values were: 553,801 W/m^2 in 2016, 561,828 W/m^2 in 2017, and 635,253 W/m^2 in 2018. In the TMY dataset, the highest TSI (1040.0 W/m^2) was noted on 20 May at noon. In 2016, maximum TSI was 1182.4 W/m^2 on 16 June at 1:00 p.m. In 2017, the highest TSI (1229.5 W/m^2) was observed on 3 July at 3:00 p.m. In 2018, TSI peaked at 1139.8 W/m^2 on 12 August at 4:00 p.m.

Air temperatures measured at the GAHE outlet and the modeled temperatures are presented in Figure 4.

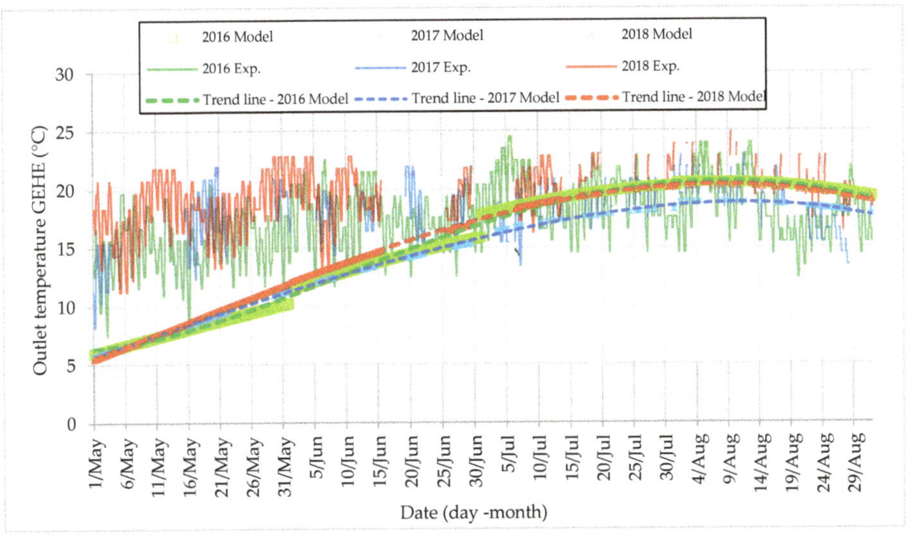

Figure 4. Temperature distribution at the GAHE outlet relative to the modeled temperatures.

The greatest differences between the experimentally measured and the modeled values were noted in the first half of the analyzed period. In May, the average monthly temperature measured at the GAHE outlet was determined at 15.2 °C in 2016, 16.7 °C in 2017, and 18.4 °C in 2018. The corresponding modeled temperatures were 8.1 °C in 2016, 8.4 °C in 2017, and 8.8 °C in 2018. In June 2016, the average monthly temperature measured at the GAHE outlet was 17.6 °C in 2016, 18.4 °C in 2017, and 19.6 °C in 2018. The corresponding data in the model were: 13.8 °C in 2016, 13.7 °C in 2017, and 14.1 °C in 2018. In July 2016, the average monthly temperature measured at the GAHE outlet was 19.2 °C in 2016, 18.6 °C in 2017, and 20.9 °C in 2018. The corresponding values in the model were 19.3 °C in 2016, 17.3 °C in 2017, and 19.1 °C in 2018. In August 2016, the average temperature measured at the GAHE outlet was 18.5 °C in 2016, 18.1 °C in 2017, and 20.8 °C in 2018. The corresponding modeled temperatures were 20.3 °C in 2016, 18.5 °C in 2017, and 19.8 °C in 2018.

Based on an analysis of the hourly differences in temperature between the GAHE inlet and the GAHE outlet (ambient air), the maximum heating load and the maximum cooling load were determined at 4.8 °C and 1.9 °C, respectively, in May 2016; 3.8 °C and 5.6 °C, respectively, in June 2016; 4.0 °C and 7.2 °C, respectively, in July 2016; 4.8 °C and 8.1 °C, respectively, in August 2016. In 2017, the maximum heating load and the maximum cooling load reached 6.4 °C and 4.8 °C, respectively in May; 3.2 °C and 4.7 °C, respectively, in June; 4.5 °C and 3.2 °C, respectively, in July; 3.5 °C and 5.6 °C, respectively, in August. In 2018, the maximum heating load and the maximum cooling load were determined at 6.5 °C and 6.8 °C, respectively in May (Table 1); 6.1 °C and 7.0 °C, respectively, in June; 4.2 °C and 6.4 °C, respectively, in July; 7.1 °C and 6.9 °C, respectively, in August.

Table 1. Distribution of average hourly temperatures measured at the GAHE inlet and outlet

Variable		Year	May	June	July	August
Hourly Ambient Air Temperature (°C)	Max.	2016	22.0	27.4	31.7	32.1
		2017	26.7	26.6	27.9	29.0
		2018	29.6	30.1	31.4	31.5
	Min.	2016	3.4	9.0	11.8	7.9
		2017	0.1	7.6	10.8	11.0
		2018	4.8	7.1	11.6	8.5
	Avg.	2016	13.4	16.8	18.9	17.5
		2017	14.2	17.3	17.9	18.3
		2018	17.4	18.6	20.7	20.2
Hourly Temperature Distribution at the GAHE Outlet (°C)	Max.	2016	20.5	22.5	24.5	24.0
		2017	21.9	21.9	21.9	22.9
		2018	22.8	22.8	23.8	24.8
	Min.	2016	7.5	11.7	14.5	12.4
		2017	8.2	14.4	13.4	13.5
		2018	11.3	13.2	16.3	16.4
	Avg.	2016	15.2	17.6	19.2	18.5
		2017	16.7	18.4	18.6	18.1
		2018	18.4	19.6	20.9	20.8
Hourly Differences in Temperature Between the GAHE Inlet and the GAHE Outlet (°C)	Heating	2016	4.8	3.8	4.0	4.8
		2017	6.4	3.2	4.5	3.5
		2018	6.5	6.1	4.2	7.1
	Cooling	2016	−1.9	−5.6	−7.2	−8.1
		2017	−4.8	−4.7	−3.2	−5.6
		2018	−6.8	−7.0	−6.4	−6.9
	Avg.	2016	1.8	0.8	0.3	1.0
		2017	1.2	0.7	0.8	0.7
		2018	1.0	0.7	0.0	0.3

Table 1. *Cont.*

Variable	Year	May	June	July	August
Total Operating Time (h)	2016	744.0	720.0	744.0	744.0
	2017	332.0	283.0	201.0	222.0
	2018	744.0	449.0	274.0	141.0
Total Air Flow (m³/h)	2016	121,344.6	112,654.7	115,274.9	107,644.8
	2017	49,690.8	40,996.5	28,225.3	32,077.2
	2018	107,924.5	62,524.7	38,119.5	19,595.9

The number of GAHE operating hours during which air was cooled deserves closer inspection. The number of hours when air was cooled by the GAHE accounted for 27% of total operating hours in 2016, 39% in 2017, and 51% in 2018. In the theoretical model with the same number of operating hours, the number of hours when air was cooled accounted for 58% of total operating hours in 2016, 73% in 2017, and 84% in 2018. A comparison of real-world data with the modeled values revealed considerable differences in GAHE performance across the years.

In Figure 5, the difference in air temperature between the GAHE inlet and outlet was plotted against the temperature of ambient air based on real-world data and modeled data. The trend lines for modeled data have a smaller slope than the trend lines for experimental data. During the conducted measurements, air was cooled only when outdoor temperature reached 15.8 °C in 2016 and 16.7 °C in successive years, whereas in the theoretical model, air was cooled already at an ambient temperature of 5.5 °C in 2016 and 2018, and 5.1 °C in 2017. Based on the measured data, air ceased to be preheated by the GAHE when ambient temperature exceeded 22.3 °C in 2016, 20.9 °C in 2017, and 22.6 °C in 2018. According to the model, the GAHE ceased to preheat air when outdoor temperature was higher than 19.2 °C in 2016, 16.9 °C in 2017, and 18.6 °C in 2018.

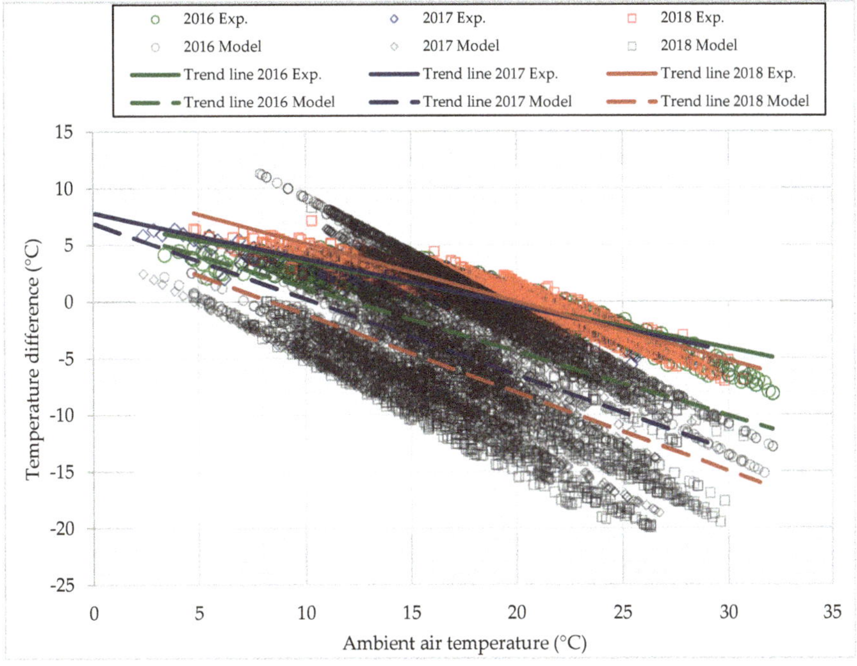

Figure 5. Differences in air temperature between the GAHE inlet and outlet (real-world and modeled data) relative to ambient air temperature.

The model produced a better fit to the experimental data in an analysis of monthly energy gain per 1 m³/h of flowing air (Figure 6). In general, heat gain was higher in the experiment than in the theoretical model. In all years of the study, the heat gain determined in the experiment was 98% higher in May and 77% higher in June than that calculated in the theoretical model. In July 2017 and 2018, heat gain was only around 30% higher in the experiment than in the model. In the remaining four months, theoretical heat gain was higher than the experimentally determined heat gain. In July 2016, August 2016, and August 2017, the modeled heat gain was approximately twice higher than that determined in the experiment.

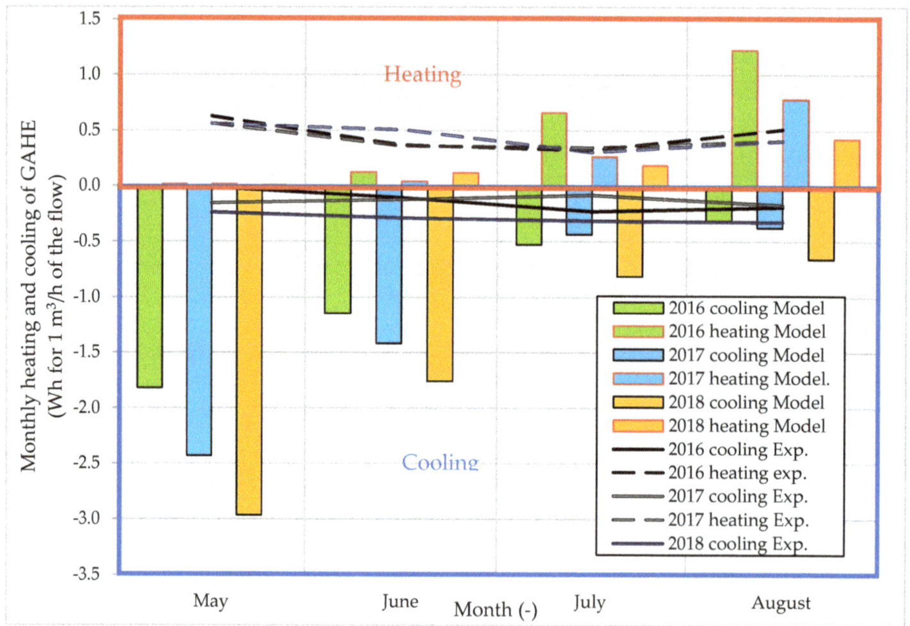

Figure 6. Monthly heating and cooling loads of the GAHE (Wh per 1 m³/h of flowing air).

The monthly cooling load (Figure 6) was clearly higher in the model than in the experiment. The difference between modeled and experimental data decreased gradually in successive months of each analyzed year. The theoretical cooling load was 130 times higher than the experimentally determined cooling load in May 2016. The modeled values were 11-fold higher than the experimental values in June 2016, and twice higher in August 2016. In 2017, the theoretical load was 16 times higher in May, 11 times higher in June, 5 times higher in July, and 2 times higher in August than that measured in the experiment. The theoretical and the experimental values were most consistent in 2018 when the modeled cooling load exceeded the experimentally determined load 12-fold in May, 3-fold in June, and 2-fold in July and August.

An analysis of hourly heating and cooling loads of the GAHE is presented in Table 2. The number of operating hours differed across the analyzed years. The GAHE operated continuously (2952 h) in 2016 and intermittently in the remaining years of the experiment. The total number of operating hours was 1038 in 2017 and 1608 in 2018. Based on the experimental data for 2016, the highest hourly heating and cooling loads were determined at 259.9 Wh and −108.6 Wh, respectively, in May; 204.0 Wh and −317.7 Wh, respectively, in June; 211.6 Wh and −401.9 Wh, respectively, in July; 231.3 Wh and −409.7 Wh, respectively, in August. The experimental data for 2017 revealed maximum hourly heating and cooling loads of 317.6 Wh and −240.4 Wh, respectively, in May; 172.5 Wh and −233.7 Wh, respectively, in June; 211.1 Wh and −153.9 Wh, respectively, in July; 171.7 Wh and −470.8 Wh, respectively, in August.

Based on the experimental data for 2018, the highest hourly heating and cooling loads were determined at 328.1 Wh and −326.0 Wh, respectively, in May; 294.4 Wh and −327.2 Wh, respectively, in June; 193.5 Wh and −311.6 Wh, respectively in July; 335.9 Wh and −340.1 Wh, respectively, in August.

Table 2. Distribution of hourly heating and cooling loads of the GAHE (Wh).

Heating and Cooling Loads of the GAHE (Wh)	Hours with Specific Heating and Cooling Loads of the GAHE (Number)					
	2016		2017		2018	
	Exp.	Model	Exp.	Model	Exp.	Model
−950	0	0	0	0	0	2
−900	0	0	0	0	0	13
−850	0	1	0	2	0	19
−800	0	8	0	7	0	23
−750	0	2	0	9	0	30
−700	0	20	0	6	0	47
−650	0	22	0	12	0	39
−600	0	34	0	17	0	57
−550	0	47	0	13	0	67
−500	0	70	0	25	0	65
−450	0	78	1	29	0	63
−400	4	95	0	31	0	80
−350	13	115	0	61	0	89
−300	17	133	0	66	7	104
−250	15	168	2	77	22	110
−200	30	160	14	93	66	138
−150	57	173	24	85	76	105
−100	97	206	33	79	112	109
−50	152	210	76	80	111	79
0	251	180	144	70	155	107
50	548	197	220	82	235	104
100	825	185	259	47	302	54
150	653	182	164	39	271	45
200	236	150	61	39	154	32
250	52	148	28	32	70	17
300	4	131	10	18	23	6
350	0	105	2	18	3	2
400	0	72	0	1	0	0
450	0	42	0	0	0	1
500	0	10	0	0	0	0
550	0	3	0	0	0	0

A comparison of hourly heating and cooling loads based on the experimental and the modeled data revealed a much higher number of hours with a heating load of 50–200 Wh (in particular in 2016) during the experiment. In turn, the number of hours with a cooling load below −450 Wh or a heating load above 350 Wh was higher in the theoretical model, and the above values were never exceeded in the experiment. The above values accounted for 18% of the modeled values in 2016, around 13% of the modeled values in 2017, and around 27% of the modeled values in 2018. Based on the modeled data for 2016, the highest hourly heating and cooling loads were determined at 179.6 Wh and −857.2 Wh, respectively, in May; 258.9 Wh and −841.0 Wh, respectively, in June; 399.0 Wh and −736.1 Wh, respectively, in July; 591.9 Wh and −587.9 Wh, respectively in August. The modeled data for 2017 revealed that the highest hourly heating and cooling loads reached 164.8 Wh and −864.9 Wh, respectively, in May; 115.4 Wh and −617.3 Wh, respectively in June; 275.0 Wh and −333.6 Wh, respectively, in July; 351.5 Wh and −460.8 Wh, respectively, in August. Based on the modeled data for 2018, the highest hourly heating and cooling loads were determined at 82.0 Wh and −981.0 Wh, respectively in May; 296.5 Wh and −822.0 Wh, respectively, in June; 290.2 Wh and −477.8 Wh, respectively in July; 432.2 Wh and −518.1 Wh, respectively, in August.

The cooling load of a GAHE is also related to moisture condensation which is calculated with the use of formulas (8) and (9). A comparison of sensible and latent cooling capacity in 2018 is presented in Table 3.

Table 3. Total cooling capacity, sensible cooling capacity, and latent cooling capacity of the GAHE in 2018

		Sensible Cooling Capacity (Wh)		Latent Cooling Capacity (Wh)		Total Cooling Capacity (Wh)	
		Hourly	Daily	Hourly	Daily	Hourly	Daily
May	Avg.	−110.8	−832.6	−80.7	−604.3	−129.1	−1649.7
	Max.	−1.0	0.0	−0.7	0.0	−0.7	−17.5
	Min.	−326.0	−3056.7	−378.6	−2335.4	−564.1	−5035.4
	Total	−25,810.9		−18,731.9		−44,542.8	
June	Avg.	−127.4	−603.0	−92.5	−419.5	−163.2	−1704.2
	Max.	−1.0	0.0	−0.8	0.0	−0.8	−12.5
	Min.	−327.2	−2625.9	−573.8	−3146.9	−661.0	−4962.7
	Total	−18,089.3		−12,586.1		−30,675.4	
July	Avg.	−97.0	−385.0	−134.1	−558.1	−173.0	−1827.1
	Max.	−1.9	0.0	−0.9	0.0	−2.4	−331.0
	Min.	−311.6	−1445.5	−660.6	−4132.0	−903.6	−5577.6
	Total	−11,934.1		−17,299.7		−29,233.9	
August	Avg.	−135.0	−204.7	−145.2	−154.6	−218.4	−928.1
	Max.	−1.9	0.0	−5.6	0.0	−14.6	−50.0
	Min.	−340.1	−1128.5	−432.3	−1475.8	−624.6	−2105.0
	Total	−6345.6		−4791.4		−11,137.0	

In 2018, sensible cooling capacity and latent cooling capacity were estimated at −25.8 kWh and −18.7 kWh, respectively, in May; −18.8 kWh and −12.6 kWh, respectively, in June; −11.9 kWh and −17.3 kWh, respectively, in July; −6.3 kWh and −4.8 kWh, respectively, in August. Total cooling capacity reached −115.6 kWh, where sensible cooling capacity accounted for 54% and latent cooling capacity – for 46% of the overall energy balance.

4. Discussion

The performance of a GAHE was analyzed between May and August of 2016, 2017, and 2018. In the studied period, air temperature, and solar irradiance continued to increase relative to TMY data as well as previous year's data. The warmest year was 2018 when the number of hours with temperatures above 25 °C was considerably higher than in the remaining years of the study. The number of hours with the lowest temperatures (below 12 °C) which generally occur in spring also continued to decrease. Sub-zero temperatures, which appear in the TMY dataset, were not encountered during the study. Although the measurements were conducted over a period of only three years which were not too distant from the period covered by the TMY dataset (1971–2000), the recorded data are indicative of climate change and global warming. The above observation is supported by the discrepancies between the experimentally measured and modeled temperatures at the GAHE outlet. The differences between the experimental and modeled values of average monthly temperature at the GAHE outlet reached 7.1 °C to 9.6 °C in May, 3.8 °C to 5.5 °C in June, 0.1 °C to 1.3 °C in July, and 0.4 °C to 1.8 °C in August. The temperatures calculated in the theoretical model were below the experimental values in the spring of each year. The modeled values were characterized by a better fit to the experimental data in summer (July and August). In the analyzed period, the GAHE was able to increase or decrease the average hourly temperature of outdoor air by up to 7.1 °C and 8.1 °C, respectively. However, the operating time of the GAHE differed in each year of the experiment. In 2016, the GAHE operated continuously, but outdoor air was precooled during only 27% of total operating time. The above can be attributed to the accumulation of heat around exchanger tubes in the ground when the GAHE remained in continuous operation mode. In the two remaining years, the GAHE operated intermittently, and outdoor air was precooled during 39% and 51% of total operating time in 2017 and 2018, respectively. The cooling load was highest in 2018 which was the warmest year and the year in which the GAHE operated intermittently. Therefore, the thermal energy content of soil was able to regenerate when the GAHE was not operating. However, according to the theoretical model, the number of hours during which outdoor air was precooled should be higher (72% on average in all years), in particular in spring, when low ground temperature after winter should support effective air cooling.

The observed discrepancies between the measured and the modeled values were also reflected in the GAHE's heating and cooling loads. According to the model, the GAHE should primarily cool outdoor air in May and June. However, the measured values indicate that air was mostly heated during that period, and the experimentally derived cooling loads were similar in all months. Both the experimental and the modeled values indicate that cooling loads should be higher in warmer years (2018) and that heating loads should be higher in colder years (2016). These observations clearly suggest that GAHEs should be popularized in an era of climate change.

Based on the conducted measurements, the annual heating load of the GAHE was determined at 210 kWh in 2016, 66 kWh in 2017, and 112 kWh in 2018. The corresponding values in the theoretical model were 224 kWh, 35 kWh, and 24 kWh. The sensible cooling load (negative values) was determined at −61 kWh in 2016, −21 kWh in 2017, and −62 kWh in 2018 based on the experimentally derived values, and at −444 kWh, −203 kWh, and −474 kWh, respectively, based on the modeled values. However, it should be noted that latent cooling capacity accounts for a high percentage of total cooling capacity (up to 50%).

5. Conclusions

The theoretical temperature at the GAHE outlet was directly influenced by the type of the model for calculating temperature distribution in the ground. Various models have been proposed in the literature. The authors selected a model based on PN-EN 16798-5 1:2017-07 to verify the extent to which this European standard is consistent with the experimentally determined temperature at the GAHE outlet in northeastern Poland.

The theoretical mathematical model applied in this study is easy to use, does not require specialist software, and can be implemented in early stages of designing HVAC systems in buildings. The values generated by the theoretical model based on Standard PN-EN 16798-5 1:2017-07 were relatively consistent with the experimentally measured data in summer, whereas considerable discrepancies between the modeled values and real-life measurements were observed in the transitional period (spring).

Further research is needed to verify the results of this study. The performance of the GAHE in fall and winter will be analyzed in an upcoming study. The volume of the experimentally measured and simulated data exceeds manuscript length limits; therefore, fall/winter data will be presented in a separate article. The performance of the GAHE will be analyzed in all months of the year, and the results will be used to optimize the theoretical mathematical model proposed by Standard PN-EN 16798-5 1:2017-07 to obtain a better fit with the experimental data.

Funding: This research received no external funding.

Conflicts of Interest: The authors declare no conflicts of interest.

References

1. Carlucci, S.; Cattarin, G.; Pagliano, L.; Pietrobon, M. Optimization of the Installation of an Earth-to-Air Heat Exchanger and Detailed Design of a Dedicated Experimental Set-up. In *Applied Mechanics and Materials*; Trans Tech Publications: Switzerland, 2014; Volume 501, pp. 2158–2161.
2. Choudhury, T.; Misra, A.K. Minimizing changing climate impact on buildings using easily and economically feasible earth to air heat exchanger technique. *Mitig. Adapt. Strateg. Glob. Chang.* **2014**, *19*, 947–954. [CrossRef]
3. Tan, L.; Love, J. A literature review on heating of ventilation air with large diameter earth tubes in cold climates. *Energies* **2013**, *6*, 3734–3743. [CrossRef]
4. Skotnicka-Siepsiak, A.; Wesołowski, M.; Piechocki, J. Validating Models for Calculating the Efficiency of Earth-to-Air Heat Exchangers Based on Laboratory Data for Fall and Winter 2016 in Northeastern Poland. *Pol. J. Environ. Stud.* **2019**, *28*. [CrossRef]
5. PN-EN 15241:2011. *Wentylacja Budynków—Metody Obliczania Strat Energii w Budynkach Spowodowanych Wentylacją i Infiltracją Powietrza*; PKN: Warszawa, Poland, 2011.
6. Popiel, C.O.; Wojtkowiak, J. Temperature distributions of ground in the urban region of Poznan City. *Exp. Therm. Fluid Sci.* **2013**, *51*, 135–148. [CrossRef]
7. Morshed, W.; Leso, L.; Conti, L.; Rossi, G.; Simonini, S.; Barbari, M. Cooling performance of earth-to-air heat exchangers applied to a poultry barn in semi-desert areas of south Iraq. *Int. J. Agric. Biol. Eng.* **2018**, *11*, 47–53. [CrossRef]
8. Sobti, J.; Singh, S.K. Earth-air heat exchanger as a green retrofit for Chandīgarh—A critical review. *Geotherm. Energy* **2015**, *3*, 1–9. [CrossRef]
9. Sehli, A.; Hasni, A.; Tamali, M. The potential of earth-air heat exchangers for low energy cooling of buildings in South Algeria. *Energy Procedia* **2012**, *18*, 496–506. [CrossRef]
10. Wengang, H.; Yanhua, L.; Mingxin, L. Application of earth-air heat exchanger cooling technology in an office building in Jinan city. *Energy Procedia* **2019**, *158*, 6105–6111. [CrossRef]
11. Ozgener, L. A review on the experimental and analytical analysis of earth to air heat exchanger (EAHE) systems in Turkey. *Renew. Sustain. Energy Rev.* **2011**, *15*, 4483–4490. [CrossRef]
12. Estrada, E.; Labat, M.; Lorente, S.; Rocha, L.A. The impact of latent heat exchanges on the design of earth air heat exchangers. *Appl. Therm. Eng.* **2018**, *129*, 306–317. [CrossRef]
13. Bordoloi, N.; Sharma, A.; Nautiyal, H.; Goel, V. An intense review on the latest advancements of Earth Air Heat Exchangers. *Renew. Sustain. Energy Rev.* **2018**, *89*, 261–280. [CrossRef]
14. Agrawal, K.K.; Misra, R.; Agrawal, G.D.; Bhardwaj, M.; Jamuwa, D.K. The state of art on the applications, technology integration, and latest research trends of earth-air-heat exchanger system. *Geothermics* **2019**, *82*, 34–50. [CrossRef]
15. Soltani, M.; Kashkooli, F.M.; Dehghani-Sanij, A.R.; Kazemi, A.R.; Bordbar, N.; Farshchi, M.J.; Elmi, K.; Gharali, M.; Dusseault, M.B. A comprehensive study of geothermal heating and cooling systems. *Sustain. Cities Soc.* **2019**, *44*, 793–818. [CrossRef]
16. Misra, A.K.; Gupta, M.; Lather, M.; Garg, H. Design and performance evaluation of low cost Earth to air heat exchanger model suitable for small buildings in arid and semi arid regions. *KSCE J. Civ. Eng.* **2015**, *19*, 853–856. [CrossRef]
17. Baglivo, C.; D'Agostino, D.; Congedo, P. Design of a ventilation system coupled with a horizontal air-ground heat exchanger (HAGHE) for a residential building in a warm climate. *Energies* **2018**, *11*, 2122. [CrossRef]
18. Wang, S.H.; Sheng, C.; Chou, H.M.; Corado, E.J.T. A Study of Solar Panel Chimney for House Ventilation. In *Applied Mechanics and Materials*; Trans Tech Publications: Switzerland, 2013; Volume 422, pp. 118–122.
19. Liu, Q.; Du, Z.; Fan, Y. Heat and Mass Transfer Behavior Prediction and Thermal Performance Analysis of Earth-to-Air Heat Exchanger by Finite Volume Method. *Energies* **2018**, *11*, 1542. [CrossRef]
20. Congedo, P.; Lorusso, C.; De Giorgi, M.; Marti, R.; D'Agostino, D. Horizontal air-ground heat exchanger performance and humidity simulation by computational fluid dynamic analysis. *Energies* **2016**, *9*, 930. [CrossRef]
21. Agrawal, K.K.; Bhardwaj, M.; Misra, R.; Agrawal, G.D.; Bansal, V. Optimization of operating parameters of earth air tunnel heat exchanger for space cooling: Taguchi method approach. *Geotherm. Energy* **2018**, *6*, 10. [CrossRef]

22. Kumar, R.; Sinha, A.R.; Singh, B.K.; Modhukalya, U. A design optimization tool of earth-to-air heat exchanger using a genetic algorithm. *Renew. Energy* **2008**, *33*, 2282–2288. [CrossRef]

23. Ahmed, S.F.; Amanullah, M.T.O.; Khan, M.M.K.; Rasul, M.G.; Hassan, N.M.S. Parametric study on thermal performance of horizontal earth pipe cooling system in summer. *Energy Convers. Manag.* **2016**, *114*, 324–337. [CrossRef]

24. Rodrigues, M.K.; da Silva Brum, R.; Vaz, J.; Rocha, L.A.O.; dos Santos, E.D.; Isoldi, L.A. Numerical investigation about the improvement of the thermal potential of an Earth-Air Heat Exchanger (EAHE) employing the Constructal Design method. *Renew. Energy* **2015**, *80*, 538–551. [CrossRef]

25. Bisoniya, T.S. Design of earth–air heat exchanger system. *Geotherm. Energy* **2015**, *3*, 18. [CrossRef]

26. Rouag, A.; Benchabane, A.; Mehdid, C.E. Thermal design of Earth-to-Air Heat Exchanger. Part I a new transient semi-analytical model for determining soil temperature. *J. Clean. Prod.* **2018**, *182*, 538–544. [CrossRef]

27. Mehdid, C.E.; Benchabane, A.; Rouag, A.; Moummi, N.; Melhegueg, M.A.; Moummi, A.; Benabdi MLBrima, A. Thermal design of Earth-to-air heat exchanger. Part II a new transient semi-analytical model and experimental validation for estimating air temperature. *J. Clean. Prod.* **2018**, *198*, 1536–1544. [CrossRef]

28. Chiesa, G. EAHX–Earth-to-air heat exchanger: Simplified method and KPI for early building design phases. *Build. Environ.* **2018**, *144*, 142–158. [CrossRef]

29. Peretti, C.; Zarrella, A.; De Carli, M.; Zecchin, R. The design and environmental evaluation of earth-to-air heat exchangers (EAHE). A literature review. *Renew. Sustain. Energy Rev.* **2013**, *28*, 107–116. [CrossRef]

30. Kusuda, T.; Achenbach, P.R. *Earth Temperature and Thermal Diffusivity at Selected Stations in the United States*; (No. NBS-8972); National Bureau of Standards: Gaithersburg, MD, USA, 1965.

31. Phetteplace, G.; Bahnfleth, D.; Mildenstein, P.; Overgaard, J.; Rafferty, K.; Wade, D.W. *District Heating Guide*; ASHRAE: Atlanta, GA, USA, 2013.

32. Tittelein, P.; Achard, G.; Wurtz, E. Modelling earth-to-air heat exchanger behaviour with the convolutive response factors method. *Appl. Energy* **2009**, *86*, 1683–1691. [CrossRef]

33. Badache, M.; Eslami-Nejad, P.; Ouzzane, M.; Aidoun, Z.; Lamarche, L. A new modeling approach for improved ground temperature profile determination. *Renew. Energy* **2016**, *85*, 436–444. [CrossRef]

34. Ozgener, O.; Ozgener, L.; Tester, J.W. A practical approach to predict soil temperature variations for geothermal (ground) heat exchangers applications. *Int. J. Heat Mass Transf.* **2013**, *62*, 473–480. [CrossRef]

35. Awadukt Thermo Ground-Air Heat Exchanger Renewable Cooling Solution for Low Energy Buildings. Available online: https://www.rehau.com/download/1420434/sales-brochure.pdf (accessed on 24 March 2020).

36. Liu, Z.; Yu, Z.J.; Yang, T.; Li, S.; El Mankibi, M.; Roccamena, L.; Qi, D.; Zhang, G. Designing and evaluating a new earth-to-air heat exchanger system in hot summer and cold winter areas. *Energy Procedia* **2019**, *158*, 6087–6092. [CrossRef]

37. Li, H.; Ni, L.; Yao, Y.; Sun, C. Experimental investigation on the cooling performance of an Earth to Air Heat Exchanger (EAHE) equipped with an irrigation system to adjust soil moisture. *Energy Build.* **2019**, *196*, 280–292. [CrossRef]

Article

Energy Analysis of a Dual-Source Heat Pump Coupled with Phase Change Materials

Michele Bottarelli [1],* and Francisco Javier González Gallero [2]

[1] Department of Architecture, University of Ferrara, Via Quartieri 8, 44121 Ferrara, Italy
[2] Escuela Politécnica Superior de Algeciras, University of Cádiz, Avenida Ramón Puyol, s/n, 11202 Algeciras, Spain; javier.gallero@uca.es
* Correspondence: michele.bottarelli@unife.it

Received: 11 May 2020; Accepted: 5 June 2020; Published: 8 June 2020

Abstract: Installation costs of ground heat exchangers (GHEs) make the technology based on ground-coupled heat pumps (GCHPs) less competitive than air source heat pumps for space heating and cooling in mild climates. A smart solution is the dual source heat pump (DSHP) which switches between the air and ground to reduce frosting issues and save the system against extreme temperatures affecting air-mode. This work analyses the coupling of DSHP with a flat-panel (FP) horizontal GHE (HGHE) and a mixture of sand and phase change materials (PCMs). From numerical simulations and considering the energy demand of a real building in Northern Italy, different combinations of heat pumps (HPs) and trench backfill material were compared. The results show that PCMs always improve the performance of the systems, allowing a further reduction of the size of the geothermal facility. Annual average heat flux at FP is four times higher when coupled with the DSHP system, due to the lower exploitation. Furthermore, the enhanced dual systems are able to perform well during extreme weather conditions for which a sole air source heat pump (ASHP) system would be unable either to work or perform efficiently. Thus, the DSHP and HGHE with PCMs are robust and resilient alternatives for air conditioning.

Keywords: shallow geothermal system; dual source heat pump; phase change materials; numerical simulations

---

1. Introduction

Ground-coupled heat pumps (GCHPs) are among the general renewable technologies used for space heating, cooling and hot water supply for buildings, especially in cold climates. Significant reductions in energy consumption, 30–70% in heating mode and 20–50% in cooling mode, have been reported when compared with conventional air-conditioning systems [1]. The ground heat exchanger (GHE), installed vertically or horizontally to exchange heat with the soil, is one of the main components of the GCHP system. Although GCHPs are more efficient than conventional air-to-water heat pumps [2], the investment cost of the GHE is high and make GCHPs less competitive than the more widespread air source heat pumps (ASHPs), especially in mild climates. Thus, despite having great potential in the future for many countries [3], the presence of GCHPs in countries with mild climates is currently not very widespread. Some novel and shallow horizontal GHE (HGHE) solutions take advantage of a better heat transfer from advanced shapes, achieving an energy performance similar to that obtained by the more thermally stable, though more expensive, vertical configurations. In this regard, new developments of very shallow geothermal systems combined with different natural and cheap backfilling materials are currently under assessment at several sites around Europe [4]. However, it seems HGHEs hold some drawbacks regarding land use [5] and payback still remains too long to justify the initial investment [6]. Yet, given the existing potential and resources, a major development is expected on shallow geothermal for HVAC (heating, ventilation and air conditioning) and GCHP systems [7–9]. In this context, the use of phase change materials (PCMs) may also assist GCHP technology in the search for making the systems more energetically efficient and economically viable. Thus, Spitler and Bernier [10] stated that the application of PCMs inside boreholes could be a viable means to reduce borehole length at peak conditions. Due to its higher thermal capacity and latent heat release during phase change, PCM is able to improve the energy storage performance of the GHE and slow down the temperature change of soil. On that basis, Eslami-nejad and Bernier [11] designed a ring with a mixture of PCM and soil around the borehole, allowing a reduction in borehole length of up to 9%. From laboratory-scale experiments and numerical models, several authors [12–14] have shown that the thermal influence radius of the GHE, and consequently the land area needed for boreholes, can be decreased by utilising PCMs as backfill material. The effect of using PCMs as grout on the thermal performance of GHEs on their own or coupled with the heat pump have also been analysed. Results provided by several studies [15,16] showed that a proper mixture of PCM and soil as grouting material of the GHE was able to smooth the thermal wave generated by the heat pump on the ground, and to enhance its coefficient of performance (COP). Some authors [17,18] have pointed out that, in hybrid systems such as dual source heat pumps (DSHPs), which change between ground and air as heat sources, the use of an air heat exchanger allows a further reduction of the GHE size, lowering the total cost of the GCHP system. Furthermore, DSHPs can be optimised by switching to the more favourable source/sink between the air and ground according to their temperature, achieving higher efficiencies in comparison with ASHPs and GCHPs [19–21]. This solution may also avoid frosting problems and save the system against extreme temperatures that affect air-mode by using the ground as an alternative heat source. Due to the spread of heat pump (HP) systems and the incontrovertible climate change, the resilience of HVAC systems to adverse weather conditions is a key factor in ensuring internal comfort.

Considering the potential of DSHPs and the advantages of using PCMs in GHEs, the present work numerically analyses the coupling of a DSHP and a novel flat-panel HGHE [21] with a mixture of sand and PCMs as backfill material into the trench. This combination, used both for space heating and cooling, has not been considered or analysed previously in the scientific literature. Thus, one of the main objectives has been to check the suitability of this system from energy numerical simulations by comparison with different possible combinations. First, the methodology proposed to assess system thermal performance is described. Model hypotheses, boundary conditions and the study cases used for comparison are shown in this section. Secondly, the results from the numerical simulations are discussed. Finally, the main conclusions of the study are drawn.

2. Methodology

This section describes the methodology followed for the aim of analysing the benefits of the DSHP coupled with a flat-panel HGHE with a sand-PCM mixture. Basically, by using numerical simulations, the energy performance of the system was compared with the one obtained by other different combinations under the same operating conditions. Thus, taking the GCHP as a reference, these combinations were set up from the inclusion or not of the air source and the sand-PCM mixture. Furthermore, the effects of varying the trench width and the energy load factor of the exchanger, estimated from the energy simulation of a real facility, were also checked.

2.1. Description of the Dual System

Although a detailed description of the dual system is given elsewhere (see [21]), its main characteristics are summarised here. The dual source heat pump combines in parallel connection an air source, through a supplementary finned tube air heat exchanger, and a ground source, by using a flat-panel HGHE (Figure 1). Three-way valves are connected at their inlets and outlets in order to switch the flow on or off. This switching between air and ground heat exchangers, which depends on source temperatures and trigger thresholds, has been implemented virtually in the numerical model of the system.

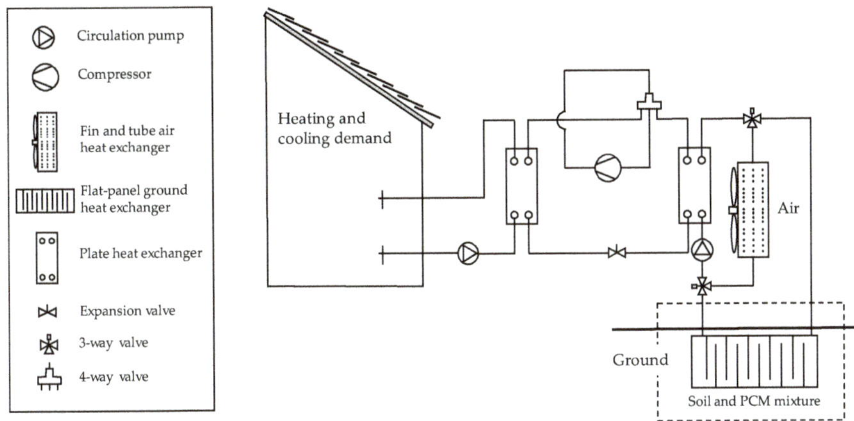

Figure 1. Layout of the dual source heat pump (DSHP) system coupled with a flat-panel horizontal ground heat exchanger (HGHE).

The flat shape of the flat-panel solution allows an easier coupling with PCMs than piping solutions of other HGHEs, such as straight pipes, slinky coils or baskets. The panel was 1.5 m high, providing a total heat transfer surface of 3 m^2 per metre trench length. The trench, which was 2.8 m deep and 0.6 m wide, was filled with a mixture of wet sand and two different PCMs, whilst the surrounding soil was assumed as wet clay. Considering cooling and heating modes, the PCMs used were paraffin A8 (PCM1) and paraffin A24 (PCM2) [22], respectively. According to their manufacturer [22], all these organic PCMs are comprised of long chain molecules, usually with a carbon backbone. The longer the chain, the higher the melting point. They were selected according to the meteorological conditions, specifically those related to minimum and maximum temperatures, in the site where heating and cooling demand were set, and ground temperatures trends of preliminary simulations. Their main thermal properties are shown in Table 1, as well as those proposed elsewhere [23] for wet sand (trench) and wet clay (surrounding soil). Volume ratios of PCM1 and PCM2 and sand into the trench were calculated considering a sand porosity reference (40%), and according to the gross building energy demand for heating (2/3) and cooling (1/3) throughout the year for the location studied.

Table 1. Physical properties of backfill materials (within trench) and surrounding ground. PCM = phase change material.

Material	Density	Thermal Conductivity	Specific Heat Capacity	Latent Heat Capacity	Melting Point
	(kg/m^3)	(W/mK)	(J/kgK)	(kJ/kg)	(°C)
Paraffin A8, PCM1 [22]	770	0.21	2160	180	8
Paraffin A24, PCM2 [22]	768	0.22	2220	250	24
Wet sand (trench)	2100	2.40	1200	–	–
Wet clay (surrounding ground)	2100	1.80	1150	–	–

2.2. Building Energy Demand and Heat Pump Efficiency

The energy requirements of the building of TekneHub Laboratory (University of Ferrara) for a one-year period were considered as the heating and cooling demand of the system. TekneHub is located in the city of Ferrara (44°50′ N, 11°37′ E) in Northern Italy. Taking this building as a reference case, EnergyPlus building energy simulation program [24] was used to estimate thermal energy loads (q_t, W/m^3) every hour for the year 2015 (Figure 2), by using local weather data collected from a meteorological station operating at TekneHub. The building envelope is well isolated according to recent Italian regulations on the energy performance of buildings. The air conditioning system consists of two air-to-air rooftop heat pumps with a capacity of 40 kW each. Further details of the building components and load calculation can be found in a study by Bottarelli and colleagues [25].

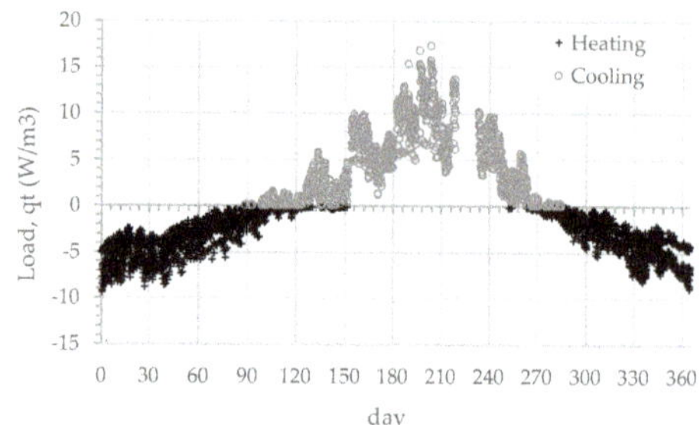

Figure 2. Heating and cooling loads per unit volume (q_t, W/m^3) of reference building space for the year 2015.

Regarding the heat pump, heating and cooling efficiencies were measured by the coefficient of performance (COP) and the energy efficiency ratio (EER), respectively. These ratios basically depend on the temperature difference between the condenser and evaporator, which is strictly related to the heat source temperatures for fixed indoor temperature set points (20 °C in winter, 26 °C in summer). The performance curves for the heat pump used by Nam et al. in their research [26] have been used here:

$$\text{COP} = 0.002T_e^2 + 0.0597T_e + 2.9028 \tag{1}$$

$$\text{EER} = 0.002T_c^2 - 0.2351T_c + 8.844 \tag{2}$$

For simplicity, evaporator temperature (T_e) and condenser temperature (T_c) are assumed to be equal to the temperature of the sources (air, ground heat exchanger) in the following.

2.3. Numerical Modelling and Boundary Conditions

The main aim here is to calculate the variation in the temperature distribution of the ground source, due to the extraction carried out by a flat-panel HGHE in coupling with a GCHP or a DSHP, and with or without a PCM mixture filling the trench. This kind of problem is governed by the equation of heat conduction in a solid (Equation (3)), and it has been numerically solved by using the commercial software COMSOL Multiphysics [27], in which some equations were implemented to characterise the PCM mixture according to the equivalent heat capacity method, as further detailed.

$$\rho c_p \frac{\partial T}{\partial t} = \vec{\nabla}\left(k\vec{\nabla}T\right) + Q \tag{3}$$

A two-dimensional (2D) computational domain which represents a cross section of the shallow GHE installation is considered, assuming the following: uniform ground temperature distribution along the z-axis, a small temperature change between the inlet and outlet along the exchanger, a negligible thermal resistance of the flat-panel, and no thermal stratification of the working fluid. Thus, the average temperature at the flat-panel is considered representative of the working fluid temperature. The simplification of the wall thermal resistance of the GHE could compromise a direct comparison with the air heat exchanger. However, it should be considered that the flat-panel prototype is actually made with walls 1.6 mm thick in high-density polyethylene (HDPE) with thermal conductivity around 0.48 W/mK). Therefore, its wall thermal resistance is around 0.0033 K/W, which is comparable with the fouling thermal resistance that affects an air heat exchanger, and even an order lower if recent techno-polymers (with thermal conductivity greater than 2 W/mK) are used instead of a standard HDPE.

Figure 3 depicts the domain and its boundary conditions. Due to symmetry conditions, only half of the domain is simulated. Figure 4 provides ground surface temperature at TekneHub Laboratory for 2015, which is highly correlated with air temperature. Maximum, mean and minimum temperatures values are 34 °C, 16 °C and 2 °C, respectively. This time series (T_s) has been set on the top of the domain (first-type boundary condition) instead of developing a more sophisticated energy balance equation at the ground surface (third-type boundary condition), since both methodologies lead to similar ground and GHE wall temperature values beyond 0.8 m deep, as reported by Bortoloni and co-workers [5]. A constant temperature of 16.7 °C was set at the bottom of the domain, corresponding to the annual average temperature measured (T_b).

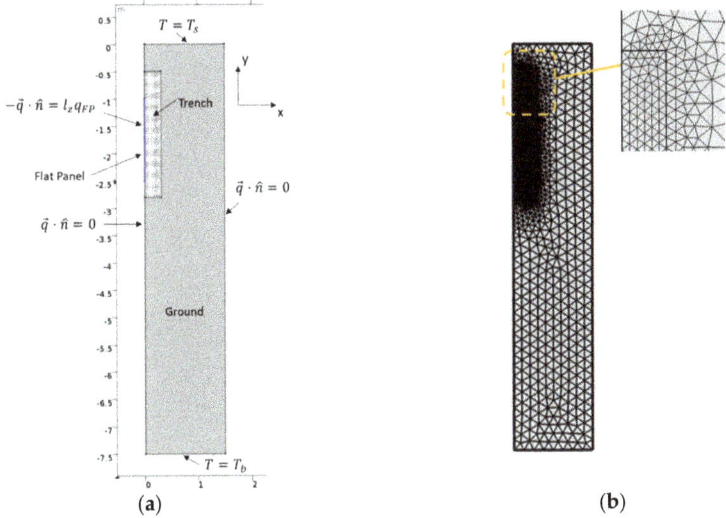

Figure 3. (a) Sketch of model domain and boundary conditions. (b) Mesh used for numerical solution.

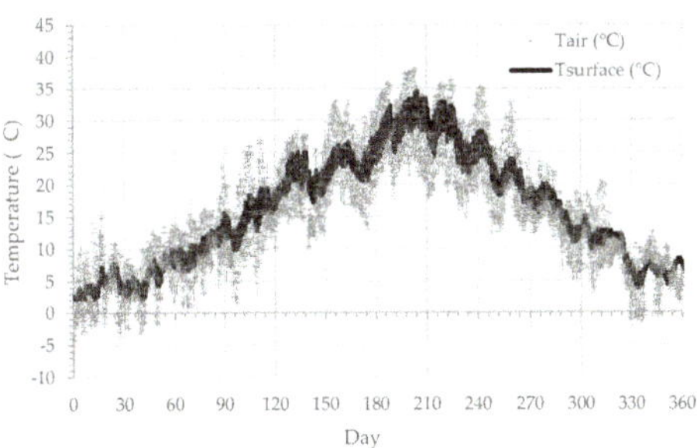

Figure 4. Time evolution of air and ground surface temperatures at TekneHub Laboratory (Ferrara, Italy, 2015).

The flat-panel GHE is simplified as a boundary heat source/sink (second-type boundary condition), for which hourly heat flux (q_{FP}, W/m^2) is calculated from Equation (4):

$$q_{FP}(t) = \frac{l_f}{S_{FP}} q_t(t) \tag{4}$$

where q_t is the thermal load per unit volume of building space (W/m^3) at time t and described in Section 2.2; S_{FP} is the heat transfer surface of a flat-panel HGHE per unit of length of the trench (2 sides times 1.5 m^2/m); l_f is the load factor calculated as the ratio between the building space volume (m^3) and the length of the trench (m). The parameter l_f provides a smart way to parametrically link the required length of the flat-panel with building energy demand (volume-to-length ratio, [21]). From the parametric study developed by Bottarelli and co-workers [21], it was concluded that the DSHP solution offered the best combination of efficiency and opportunity to reduce installation cost when l_f is between 10 m^3/m and 15 m^3/m. In the present study, l_f values varied within the range 10 m^3/m to 30 m^3/m, with 10 m^3/m and 20 m^3/m as reference values for the GCHP and DSHP, respectively. Finally, an adiabatic condition was set to the rest of the side boundaries of the domain.

The problem was solved numerically by using COMSOL Multiphysics [27]. Using the finite element method, a preliminary grid independence study was carried out. The case with the presence of PCMs and the dual source functionality was considered to evaluate the average temperature at the flat-panel in an unsteady state for a period of 40 days and six different meshes. Newton's iterative method was selected to solve the system of equations at each time-step. Absolute and relative tolerances were set to 10^{-5} °C and 10^{-3}, respectively. Figure 5 depicts all resulting values; relative changes in the estimation of average flat-panel temperature were smaller than 1%. A final mesh consisting of around 7000 triangular elements (Figure 3b), finer around the GHE trench, was finally selected to simulate all cases.

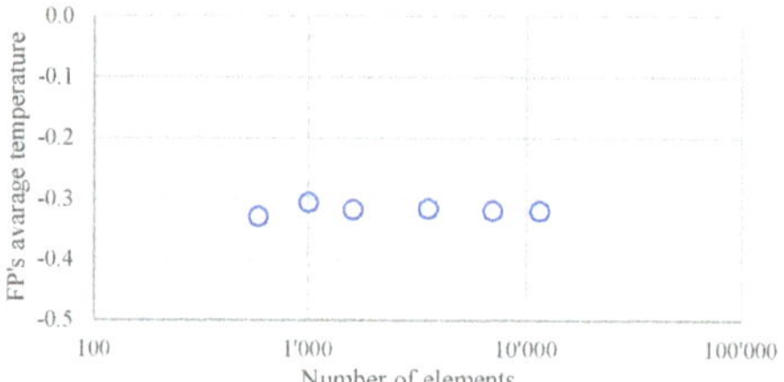

Figure 5. Results of mesh independence study.

When the mixture of sand and PCMs is considered within the trench, heat transfer is simplified to heat conduction within an equivalent solid mixture, according to the equivalent heat capacity method. Thus, the equivalent mixture density and thermal conductivity are calculated as volume weighted averages according to component volume proportions, r_i (Equations (5) and (6)). In contrast with other studies [16,21], the specific heat of the mixture was calculated as a mass weighted averaged (Equation (7)). These mixture properties are then calculated as follows:

$$\rho_{eq} = \sum_{i=0}^{n} r_i \cdot \rho_i(T) = \left(1 - \sum_{i=1}^{n} r_i\right) \cdot \rho_s + \sum_{i=1}^{n} r_i \cdot \left[(1 - H_i(T)) \cdot \rho_i^S + H_i(T) \cdot \rho_i^L\right] \tag{5}$$

$$k_{eq} = \sum_{i=0}^{n} r_i \cdot k_i(T) = \left(1 - \sum_{i=1}^{n} r_i\right) \cdot k_s + \sum_{i=1}^{n} r_i \cdot \left[(1 - H_i(T)) \cdot k_i^S + H_i(T) \cdot k_i^L\right] \tag{6}$$

$$\begin{aligned}
c_{P_{eq}} &= \sum_{i=0}^{n} Y_i \cdot c_{P_i}(T) \\
&= \left(1 - \sum_{i=1}^{n} Y_i\right) \cdot c_{P_s} + \sum_{i=1}^{n} Y_i \\
&\quad \cdot \left[(1 - H_i(T)) \cdot \left(c_{P_i}^S + h_i^{SL} \cdot D_i(T)\right) + H_i(T) \cdot \left(c_{P_i}^L + h_i^{SL} \cdot D_i(T)\right)\right]
\end{aligned} \tag{7}$$

where r_i and Y_i are the volume and mass fractions of each i component, respectively. These ratios are related as follows:

$$Y_i = \sum_{i=0}^{n} \frac{r_i \rho_i}{\sum_{i=0}^{n} r_i \rho_i} \tag{8}$$

In the present study, two PCMs are considered, such that $n = 2$ ($i = 0$ for sand, 1 for PCM1 and 2 for PCM2). $H_i(T)$ and $D_i(T)$ are step and normalised Dirac's pulse functions, respectively, used to represent the evolution of the different physical properties during the phase change, similarly to the approach reported by Bottarelli et al. [16]. Specifically, $c_{P_i}(T)$ includes the specific heat capacity in each phase ($c_{P_i}^S$, $c_{P_i}^L$) and latent heat fusion (h_i) for PCM's i component. The distribution of latent heat is approximated by using the $D_i(T)$ function, in such a way that the total heat per unit volume released during the phase transformation equals the latent heat. A transitional interval of 8 °C centred at the melting temperature was considered. As sand porosity is not greater than 40%, the sum of the PCMs' volume ratios ($r_1 + r_2$) is limited and set to 0.40; only in a single case was this value exceeded. In addition, according to the ratio of the periods of heating and cooling demand (Figure 2), r_1 and r_2 were set equal to 0.25 and 0.15, respectively.

Furthermore, a virtual control of the boundary condition at flat-panel was programmed in COMSOL in order to simulate the switching between air and ground sources in the DSHP case, as described by Bottarelli and colleagues [21]. The flowchart in Figure 6 describes the performance routine. In winter, the ground achieves better working conditions when its temperature is higher than outdoor air temperature. Ground source is used in the model when outdoor air temperature is smaller than 5 °C ($T_{air\ limit}$), and also when ground source temperature (T_g) exceeds air temperature by an onset temperature differential, ΔT. According to the guidelines for the maximum benefits of DSHPs reported by Bottarelli and colleagues [21], the corresponding temperature differential was around 5 °C for the range of variation of l_f (from 10 to 30 in this study). Both rules ($T_{air\ limit} = 5$ °C; $\Delta T = 5$ °C) aim to preserve the ground source for the late winter, because the exploitation of the high ground temperature at the winter beginning would lead to quick depletion of the underground thermal energy storage.

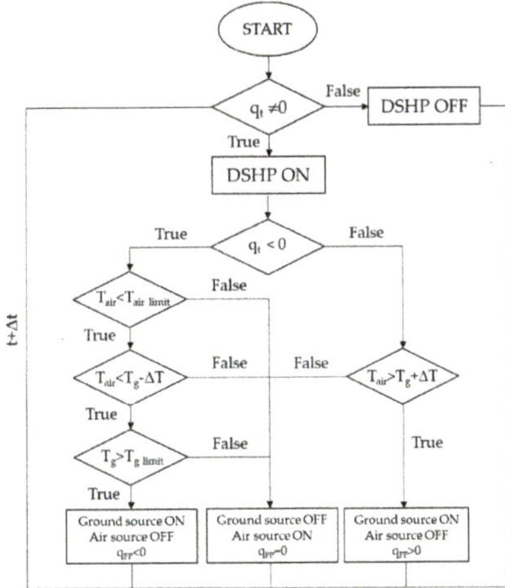

Figure 6. Flowchart of DSHP performance routine for heating and cooling mode.

An operating threshold is finally set for the ground source temperature ($T_{g\ lim} = -2$ °C) to prevent working fluid in the flat-panel from freezing. In summer, ground source is on when the difference between air and ground source temperatures is higher than the onset differential.

Air source temperature was defined as that of outdoor air (year 2015) at an hourly time scale. The temperature of the ground source was estimated by the model every time step and reported at an hourly scale as the average temperature at the flat-panel boundary.

The initial temperature distribution in the solid domain was derived from experimental measurements of ground temperature at different depths (Figure 7). However, all models were initially run for a full year to achieve an initial condition already affected by their specific yearly exploitation.

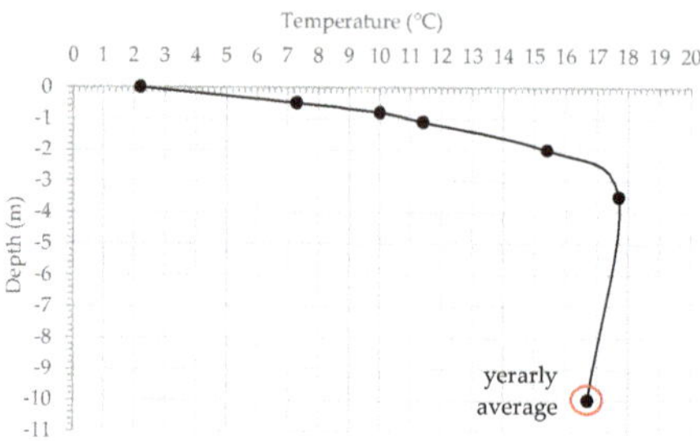

Figure 7. Initial vertical distribution of undisturbed ground temperature (TekneHub Laboratory, 1 January 2015).

2.4. Case Studies

Different case studies have been numerically simulated in unsteady heat transfer conditions in order to quantify the potential benefits of using both the dual system, specifically a DSHP coupled with the flat-panel (FP) HGHE, and a mixture of sand and PCM as backfill material. Water to water GCHP with sand as backfill material was taken as a reference case. Table 2 describes all the cases considered. Load factor (l_f) was increased in relation to the reference case (DP00), for which l_f equals 10 m^3/m (building cubic metre over metre FP length), in order to assess the energy performance and capability of the enhanced systems by using smaller heat transfer surfaces of the flat-panel and PCMs. Indeed, numerical simulations previously carried out showed that higher values of l_f for the DP00 leaded to temperatures several degrees below zero, which are not compliant with common operating conditions of this kind of facility. Furthermore, the effect of a reduction of trench width (w_t) was also analysed in case DP11#, as well as higher volume PCM ratios ($r_1 = 0.50$ and $r_2 = 0.20$) and higher thermal PCM conductivity ($k_1 = k_2 = 1.2$ W/mK) in case DP11+.

Table 2. Description of case studies.

Case	Heat Pump	Backfill Material	Load Factor, l_f (m^3/m)	Trench Width, w_t (m)
DP00	GCHP	Sand	10	0.60
DP01	GCHP	Sand and PCMs	10	0.60
DP10	DSHP	Sand	20	0.60
DP11	DSHP	Sand and PCMs	20	0.60
DP11*	DSHP	Sand and PCMs	30	0.60
DP11#	DSHP	Sand and PCMs	30	0.40
DP11+	DSHP	Sand and PCMs	20	0.60

Note: *, # cases referring to DP11 as PCMs; + case with higher PCMs thermal conductivity and mass.

In all cases, the novel flat-panel HGHE is coupled with the GCHP or DSHP. The bias between the experiments and the numerical model implemented in COMSOL of this HGHE is less than 1 °C on average, as reported by Bottarelli and colleagues [21]. When the DSHP is on, the value of the onset temperature differential (ΔT) was set to 5 °C.

3. Results and Discussion

All models were run for a two-year period by repeating weather data of the year 2015 and under the same boundary conditions, in order to set the typical initial conditions of this kind of facility and to ensure that all models reproduced their stationary trend according to their different exploitations. Thus, Figure 8a shows flat-panel temperatures for the second simulation year cases DP00, DP01, DP10 and DP11, together with outdoor air and undisturbed ground temperature (at a depth of 1.7 m). Similarly, Figure 8b shows cases DP11, DP11*, DP11# and DP11+. As it can be inferred from the series of undisturbed ground temperature, despite ground thermal exploitation and unlike deep geothermal systems, thermal drift is avoided by using this shallow HGHE.

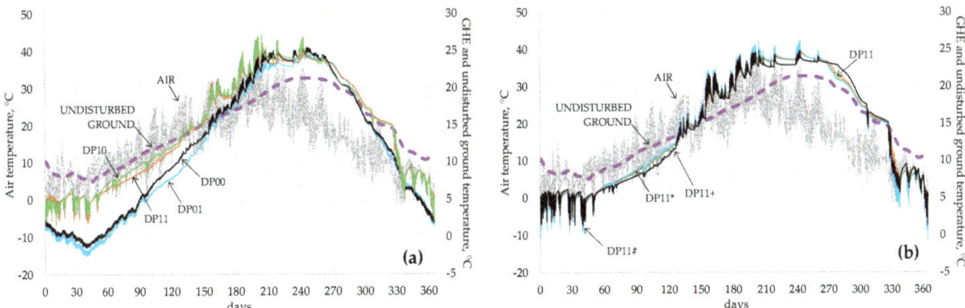

Figure 8. Annual trend of temperatures (**a**) for cases DP00, DP01, DP10, DP11 and (**b**) DP11, DP11*, DP11#, DP11+.

Flat-panel temperature for DSHP cases (DP10 and DP11) is higher than for GCHP cases (DP00 and DP01), since the dual system is able to switch to the air when ground temperature is lower in winter (Figure 8a). In this way, the lowest flat-panel temperatures for DP10 and DP11 cases are 2.1 °C and 1.7 °C, respectively. This issue may avoid the use of anti-freeze (e.g., propylene glycol) in the secondary loop of the system and reduce or exclude all frosting problems at the air heat exchanger. In contrast, the lowest flat-panel temperatures for GCHP cases DP00 and DP01 are −1.3 °C and −2.4 °C, such that anti-freeze usage is needed. As a consequence, dual strategy leaves a warmer ground at the beginning of summer which is less advantageous for this period. On the contrary, the lower temperature of GCHP systems makes the condition disadvantageous in wintertime, but more profitable in summer. When PCMs are used in the backfill material (DP01 and DP11) a similar behaviour is found, although the performance of the system is clearly lower in wintertime (minimum temperatures of −2.4 °C and 1.7 °C). These results are mainly due to the low thermal conductivity of the mixture in comparison with that of the sand. However, despite this drawback, the performance is improved in summertime, according to the higher cold energy stored in the ground, as maximum temperatures of 24.9 °C and 25.7 °C are achieved for DP01 and DP11, respectively. In Figure 8b, DP11* and DP11# minimum temperatures are −0.3 °C and −0.6 °C in wintertime, respectively, whilst 1.3 °C is that of DP11+. In summertime, the maximum temperatures are 26.1 °C, 26.2 °C and 24.7 °C for DP11*, DP11# and DP11+, respectively.

Overall, dual systems are able to perform well during extreme weather conditions (very low or very high outdoor air temperatures) for which a sole ASHP system would be unable either to work or perform efficiently. This makes the dual system a robust alternative in Southern European countries in which weather conditions are expected to become more severe in the future, with higher inter-annual increase in summer temperatures and low variability in current winter temperatures [28].

Details of the annual trend are shown at the beginning of the year in Figure 9, and by the middle of the year in Figure 10; that is, in winter and summer, respectively. When comparing, it should be noted that case DP11* has a higher l_f factor (30 m^3/m), DP11# also includes a narrow trench and consequently a smaller quantity of PCMs, whilst DP11+ has a standard trench (60 cm wide) and load factor (20 m^3/m), but uses PCMs with higher thermal conductivity.

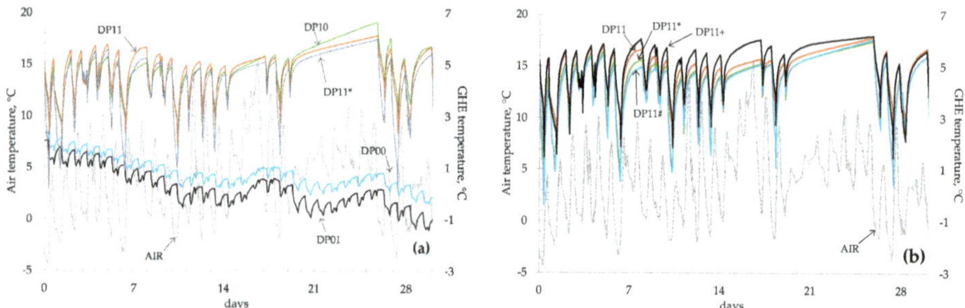

Figure 9. Details of winter temperatures (**a**) for cases DP00, DP01, DP10, DP11 and (**b**) DP11, DP11*, DP11#, DP11+.

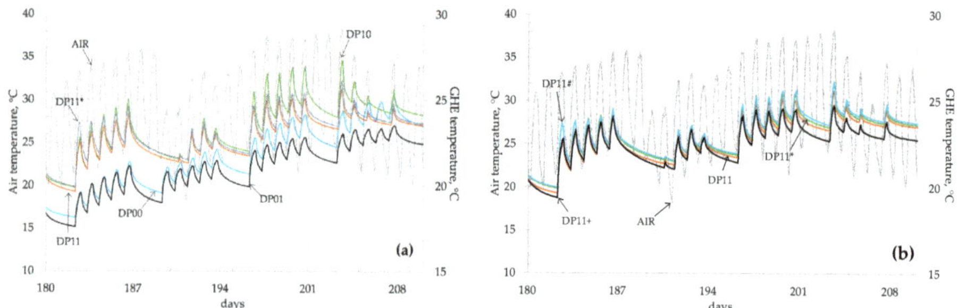

Figure 10. Details of summer temperatures (**a**) for cases DP00, DP01, DP10, DP11 and (**b**) DP11, DP11*, DP11#, DP11+.

In Figure 9a, the steady decrease of flat-panel temperature in January due to a continuous ground exploitation carried out by the GCHP (DP00 and DP01) contrasts with the nearly constant trend shown by the DSHP (DP10, DP11, DP11*). DP11+ shows the promptest celerity in recovering the exploitation in comparison with all other DSHPs (Figure 9b). Cases with PCMs show better performance than those without (Figure 10a), whilst the aforementioned behaviour of DP11+ is confirmed also in summertime (Figure 10b). Therefore, this last property seems to be the key factor in using PCMs: not to penalise the improvement of energy storage with a lower thermal conductivity of the backfill material.

In Figure 11, the resulting performance of the GCHP (DP00, DP01) and DSHP (DP10, DP11) are depicted in terms of COP (wintertime, Figure 11a) and EER (summertime, Figure 11b), according to the previous Equations (1) and (2). GCHP cases show better COP values than DSHP cases at the beginning of winter, when the ground temperature is very high due to the previous heating up in summer. However, the reverse happens late in winter, when the ground has been deeply exploited by the GCHP operation, and partially saved by the DSHP mode. This inversion does not happen in summertime (EER), since the lowest ground temperature performs better during the whole summer.

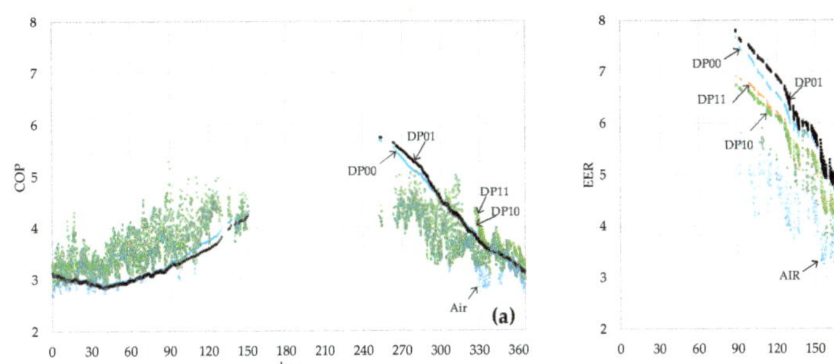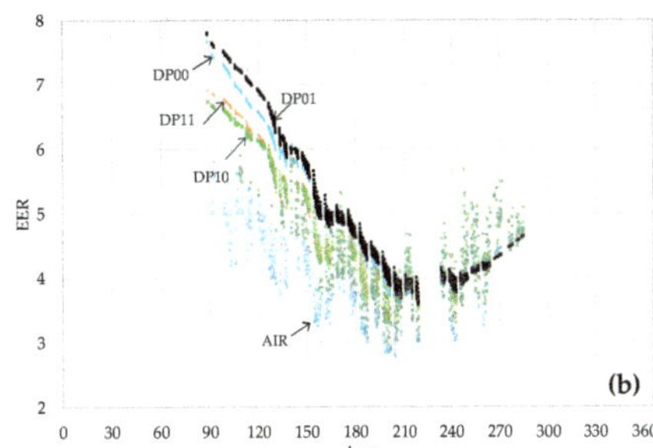

Figure 11. (**a**) Coefficient of performance (COP) and (**b**) energy efficiency ratio (EER) hourly trend.

A summary of the energy exchange and performance of the different systems in winter, summer and the whole year is shown in Tables 3–5, respectively. The values of thermal (Qt) and electrical (Qe) energy exchanged per unit length of trench (kWh/m) are given too.

Table 3. Summary of energy performance for the different case studies in wintertime.

Variable	Unit	ASHP	DP00	DP01	DP10	DP11	DP11*	DP11#	DP11+
l_f	m³/m	(10)	10	10	20	20	30	30	20
w_t	m	–	0.6	0.6	0.6	0.6	0.6	0.4	0.6
Qt_{DSHP}	kWh/m	–	–	–	−295.1	−295.1	−442.7	−442.7	−295.1
Qt_{GCHP}	kWh/m	–	−147.6	−147.6	−58.5	−60.0	−70.5	−69.7	−69.1
Qt_{ASHP}	kWh/m	−147.6	–	–	−236.6	−235.1	−372.2	−373.0	−226.1
Qe_{DSHP}	kWh/m	–	–	–	121.3	121.1	184.6	184.6	119.9
Qe_{GCHP}	kWh/m	–	65.4	65.9	23.8	24.4	29.6	29.3	27.7
Qe_{ASHP}	kWh/m	63.8	–	–	97.5	96.8	155.0	155.3	92.2
COP_{DSHP}	–	–	–	–	3.43	3.44	3.40	3.40	3.46
COP_{GCHP}	–	–	3.26	3.24	3.46	3.46	3.38	3.38	3.50
COP_{ASHP}	–	3.31	–	–	3.43	3.43	3.40	3.40	3.45
Time	h	5228	5228	5228	637	641	504	500	720
q_{FP_ave}	W/m	–	−28.2	−28.2	−91.9	−93.7	−139.9	−139.9	−95.9

Table 4. Summary of energy performance for the different case studies in summertime.

Variable	Unit	ASHP	DP00	DP01	DP10	DP11	DP11*	DP11#	DP11+
l_f	m³/m	(10)	10	10	20	20	30	30	20
w_t	m	–	0.6	0.6	0.6	0.6	0.6	0.4	0.6
Qt_{DSHP}	kWh/m	–	–	–	131.3	131.3	197.0	197.0	131.3
Qt_{GCHP}	kWh/m	–	65.7	65.7	65.3	63.3	68.5	69.7	66.9
Qt_{ASHP}	kWh/m	65.7	–	–	66.1	68.0	128.5	127.3	64.4
Qe_{DSHP}	kWh/m	–	–	–	26.3	26.4	40.6	40.6	26.2
Qe_{GCHP}	kWh/m	–	12.4	12.2	12.6	12.2	13.3	13.6	12.8
Qe_{ASHP}	kWh/m	14.2	–	–	13.7	14.2	27.3	27.0	13.4
EER_{DSHP}	–	–	–	–	3.99	3.98	3.85	3.85	4.02
EER_{GCHP}	–	–	4.31	4.37	4.18	4.20	4.14	4.13	4.24
EER_{ASHP}	–	3.61	–	–	3.81	3.79	3.70	3.71	3.80
Time	h	1304	1304	1304	608	604	494	488	635
q_{FP_ave}	W/m	–	50.4	50.4	107.3	104.8	138.6	142.8	105.4

Table 5. Annual energy performance for the different case studies. Minimum and maximum GHE temperatures are also shown.

Variable	Unit	ASHP	DP00	DP01	DP10	DP11	DP11*	DP11#	DP11+
l_f	m^3/m	(10)	10	10	20	20	30	30	20
w_t	m	–	0.6	0.6	0.6	0.6	0.6	0.4	0.6
T_{min}	°C	–	−1.3	−2.4	2.1	1.7	−0.3	−0.6	1.3
T_{max}	°C	–	26.1	24.9	27.3	25.7	26.1	26.2	24.7
Q_t	kWh/m	213.2	213.2	213.2	426.5	426.5	639.7	639.7	426.5
Q_e	kWh/m	77.7	77.7	78.2	147.6	147.5	225.2	225.2	146.1
COP/EER	–	3.37	3.42	3.42	3.53	3.53	3.48	3.48	3.56
Time	h	6532	6532	6532	1245	1245	998	988	1355
q_{FP_ave}	W/m	–	32.7	32.7	99.4	99.1	139.2	141.1	100.3

In wintertime (Table 3), DP10 and DP11 perform similarly, and better than DP00 and DP01. Specifically, DP11 shows a higher COP than DP01 (3.44 against 3.24), and also DP10 (COP value of 3.43) compared to DP00 (COP value of 3.26). Furthermore, DP11* and DP11# achieve very similar performance if compared with DP11, but with a significant reduction of the HGHE size (l_f of 30m^3/m against 20 m^3/m), and a narrower trench for case DP11# (40 cm against 60 cm).

Due to the much shorter use of the ground heat exchanger carried out by the DSHP cases (DP10, DP11, DP11*, DP11# and DP11+), very high values of heat flux per metre of trench are obtained, that range from more than three times for case DP11+ up to almost five times for case DP11*, when compared with GCHP. This is mainly related to the shorter ground exploitation of the DSHP cases, for which heat transfer takes from 500 h (DP11#) up to 720 h (DP11+).

This seems to reflect a good performance of the dual system even if coupled with a shorter HGHE, and that PCMs can improve the system performance only if their thermal conductivities are higher. Therefore, the common low thermal conductivity of PCMs attenuates their potential good performance in wintertime related to their underground thermal energy storage.

In comparison with the cheapest ASHP which shows an overall COP value around 3.31, the best value of a DSHP, around 3.46 (DP11#) does not seem to justify the investment of PCMs and GHE. However, it should be noted that the air temperature drops below 3 °C for 747 h over 5228 h of heating time, that is, 14.2% of the working period, in which frosting issues are very common at this location. As a consequence, the ASHP performance given here should be considered overestimated up to a value around 15%, as reported by Dongellini and colleagues [29].

In summertime, DP01 shows the best performance (EER value of 4.37), as justified by the ground cooling occurring for long thermal exploitation in wintertime, and the presence of PCMs. The best DSHP case is DP11# with an EER value of 4.02, resulting from the 4.24 and 3.80 values in ground and air exploitation, respectively. It is worth noting that DP11# uses a GHE which is three times shorter than the one used by case DP00. The ASHP performance is the lowest (EER value of 3.61 value) and a more interesting gap is evident, especially related to a larger exploitation of the GHE. In summer season, the overall needs of air conditioning cover a period of 1304 h; when the DSHP is used, the ground exploitation goes from a minimum of 488 h (37.4%, DP11#) to a maximum of 635 h (48.7%, DP11+). The average heat flux occurring at the flat-panel GHE (q_{FP_ave}) shows a minimum of 50.4 W per length metre of trench in GCHP cases (DP00, DP01) and a maximum of 142.8 W/m in DSHP case (DP11#), which demonstrates the high performance of the flat-panel. q_{FP_ave} values are quite comparable in winter and summer for DSHP cases, while winter values halved those in summer for GCHP cases. This difference is always attributable to the exploitation time of the ground, unchanged for DSHP cases (500–700 h) unlike GCHP cases (with more than 5000 h in wintertime against 1300 h in summertime).

As expected from the annual trend of flat-panel temperature, all DSHP cases show higher efficiencies than GCHP cases. Yet, they still perform better than the ASHP. Regarding the heat flux per metre of trench, DSHP cases provide values which are several times those given by GCHP cases.

During the whole year, DP00 and DP01 show COP/EER values of 3.42, the ASHP of 3.37 and the DSHP from 3.48 to 3.56. Overall, the average heat flux occurring at the flat-panel, coupled with a DSHP system, is three to five times higher than that of a GCHP.

Finally, an analysis about how the PCM behaves in the different cases was carried out. Thus, Figure 12 shows the solid phase time series in terms of mass fraction for the trench. In winter, a more efficient use of PCM1 is made by DP01 (Figure 12a) than by DP11 (Figure 12b). Hence, PCM1 solidifies reaching a peak of 100% (DP01), while it is of 80% in DP11. A slight improvement is achieved by DP11*, while DP11+ gets slightly lower results (Figure 12c,d). In summertime, the fraction of PCM2 that becomes liquid is very similar in all cases. This result could also suggest the advisability of using backfill materials in the trench with higher thermal conductivity.

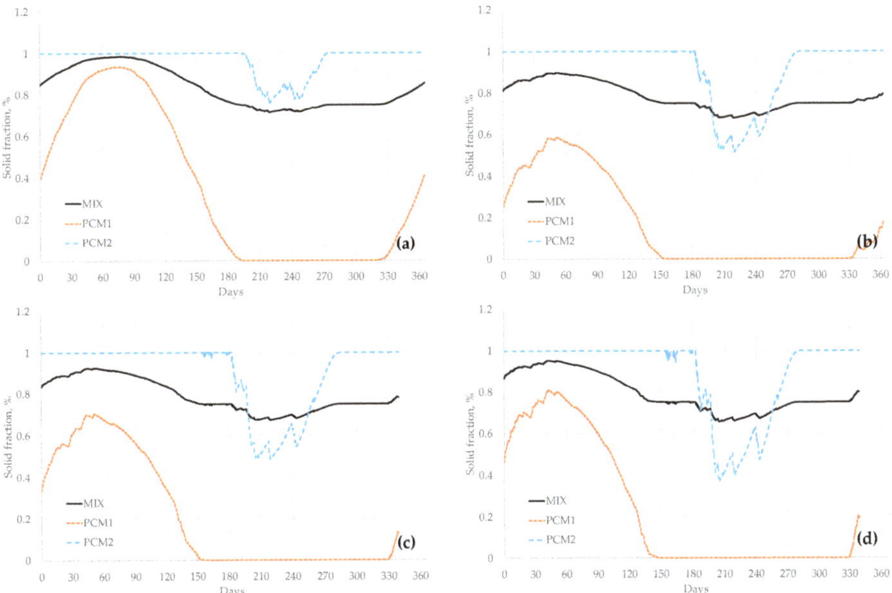

Figure 12. Solid phase fraction of the PCM components and overall mixture (MIX) in the trench. (**a**) DP01, (**b**) DP11, (**c**) DP11*, (**d**) DP11+.

To complete the results depiction, a sequence of the thermal field at 41.333 days of simulation time is presented in Figure 13 for different cases. The first two images (DP00, DP01) show the large exploitation carried out by the GCHP in comparison with the DSHP; no relevant differences are shown when PCMs are used in coupling with a GCHP. More interesting differences are detectable among DSHP cases. In the sequence DP10, DP11 and DP11*, the ground temperature rises in the domain from the bottom to the top, whilst the flat-panel temperature decreases. Temperature distribution for DP11 and DP11# cases are quite similar, while DP11+ shows the highest temperature values among all.

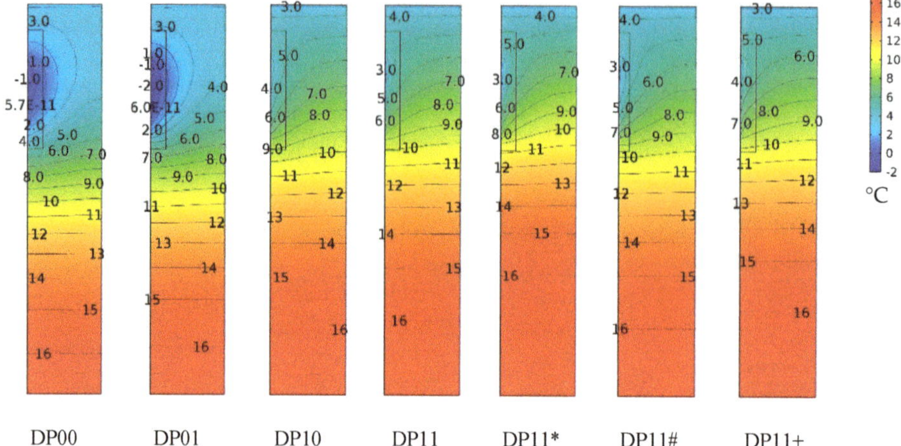

Figure 13. Thermal field at 41.333 days of simulation for cases DP00, DP01, DP10, DP11, DP11*, DP11#, DP11+.

4. Conclusions

In this study, the energy performance of the coupling of a DSHP and a novel flat-panel HGHE with a mixture of sand and PCM backfill material into the trench was analysed numerically. To the authors' knowledge, this combination, used both for space heating and cooling, has not been considered or studied previously in the scientific literature. A methodology based on energy numerical simulations, in which different source heat pumps (GCHP, ASHP and DSHP) and configurations (with and without PCMs, and different GHEs and trench sizes) were compared under the same real operating conditions, was developed to check the suitability of the system. The main conclusions of the study are as follows.

- Numerical simulations show that the overall performance of a ground source heat pump is higher, although not very significantly, than the one given by an air source heat pump.
- The use of shallow HGHEs avoids thermal drift of ground temperature and shows significant values of heat flux value per unit length of the exchanger, even when compared to vertical boreholes. The flat-panel shows an annual average heat flux of around 33 W/m when it is coupled with a GCHP, and more than 140 W/m with a DSHP.
- The dual-source functionality optimises the performance of a heat pump and drastically allows the reduction of the HGHE length to up to several times that of a full GCHP, causing a significant reduction in installation costs. Moreover, the value estimated for the lowest working fluid temperature precludes the use of glycol, with benefits in terms of costs and environmental impact.
- The use of PCMs as backfilling material can improve the overall energy performance only if a significant increase of their thermal conductivity is carried out, as the common low value of the pure paraffins' conductivity reduces the benefit of their high latent heat. Therefore, the improvement of their thermal conductivity is a key factor for their exploitation in similar applications.
- Summer thermal performance of a DSHP coupled with a flat-panel GHE and a mixture of enhanced PCMs as backfill material is better than in wintertime, according to the higher temperature difference between undisturbed ground and GHE.

Therefore, DSHP coupled with a shallow GHE backfilled with PCMs having high thermal conductivity is an interesting alternative for the air conditioning of buildings, especially in high latitudes of Southern European countries, where cooling requirements are relevant. This solution may also avoid frosting problems and save the system against extreme temperatures that affect the air-mode, by using the ground as an alternative heat source, which makes the HVAC system more resilient against the incontrovertible climate change.

A future development of this research will be the experimental validation of the main conclusions and the assessment of the economic viability of the enhanced system.

Author Contributions: All authors contributed equally to this work. All authors designed the simulations, discussed the results and implications and commented on the manuscript at all stages. M.B. conceptualized the study and methodology. F.J.G.G. carried out the formal analysis and led the development of the paper. Both authors performed numerical simulations and result analysis and discussion. All authors have read and agreed to the published version of the manuscript.

Funding: This research received no external funding.

Conflicts of Interest: The authors declare no conflict of interest.

Nomenclature

c_P	Specific heat capacity, J/(kg $\cdot$ K)
$c_{P_{eq}}$	Mixture specific heat capacity, J/(kg $\cdot$ K)
c_{P_i}	Specific heat capacity of component i, J/(kg $\cdot$ K)
c_{P_s}	Specific heat capacity of wet sand, J/(kg $\cdot$ K)
$c_{P_i}^s$	Specific heat capacity of component i in solid phase, J/(kg $\cdot$ K)
$c_{P_i}^L$	Specific heat capacity of component i in liquid phase, J/(kg $\cdot$ K)
D_i	Dirac's pulse function, K^{-1}
h_i^{SL}	Latent heat of fusion, J/kg
H_i	Step function
k	Solid thermal conductivity, W/(m $\cdot$ K)
k_{eq}	Mixture thermal conductivity, W/(m $\cdot$ K)
k_i	Thermal conductivity of component i, W/(m $\cdot$ K)
k_s	Thermal conductivity of wet sand, W/(m $\cdot$ K)
k_i^s	Thermal conductivity of component i in solid phase, W/(m $\cdot$ K)
k_i^L	Thermal conductivity of component i in liquid phase, W/(m $\cdot$ K)
l_z	Flat-panel height, m
$\hat{n}$	Normal unit vector
q	Heat load for unit building volume, W/m^3
q_{FP}	Heat flux at flat-panel, W/m^2
q_t	Energy load per unit volume of building space at time t, W/m^3
Q	Heat source/sink, W/m^3
Q_e	Thermal energy exchanged per unit length of trench, kWh/m
Q_t	Electrical energy exchanged per unit length of trench, kWh/m
l_f	Load factor, m^3/m
r_i	Volume fraction of i component, m^3 component i/m^3 mixture
S_{FP}	Flat-panel heat transfer surface per unit of trench length, m^2/m
t	Time, s
T	Temperature, K
T_{air}	Air temperature, K
$T_{air\ limit}$	Limiting air temperature, K
T_b	Temperature at the bottom of the domain, K
T_c	Condenser temperature, K
T_e	Evaporator temperature, K
T_{FP}	Average Flat-panel temperature, K
T_g	Ground temperature, K
$T_{g\ limit}$	Limiting ground temperature, K
T_s	Temperature at ground surface, K
w_t	Trench width, m
Y_i	Mass fraction of component i

Greek Symbols

ρ_{eq}	Mixture density, kg/m^3
ρ_i	Density of component i, kg/m^3
ρ_s	Density of wet sand, kg/m^3
ρ_i^s	Density of component i in solid phase, kg/m^3
ρ_i^L	Density of component i in liquid phase, kg/m^3

Acronyms

ASHP	Air source heat pump
DSHP	Dual source heat pump
COP	Coefficient of performance
EER	Energy efficiency ratio
FP	Flat-panel
GHE	Ground heat exchanger
GCHP	Ground-coupled heat pump
HGHE	Horizontal ground heat exchanger
PCM	Phase change material

References

1. Benli, H.; Durmus, A. Evaluation of ground-source heat pump combined latent heat storage system performance in greenhouse heating. *Energy Build.* **2009**, *41*, 220–228. [CrossRef]
2. Urchueguía, J.F.; Zacarés, M.; Corberán, J.M. Comparison between the energy performance of a ground coupled water to water heat pump system and an air to water heat pump system for heating and cooling in typical conditions of the European Mediterranean coast. *Energy Convers. Manag.* **2008**, *49*, 2917–2923. [CrossRef]
3. Bayer, P.; Saner, D.; Bolay, S.; Rybach, L.; Blum, P. Greenhouse gas emission savings of ground source heat pump systems in Europe: A review. *Renew. Sust. Energy Rev.* **2012**, *16*, 1256–1267. [CrossRef]
4. Bertermann, D.; Bernardi, A.; Pockelé, L.; Galgaro, A.; Cultrera, M.; de Carli, M.; Müller, J. European project "Cheap-GSHPs": Installation and monitoring of newly designed helicoidal ground source heat exchangeron the German test site. *Environ. Earth Sci.* **2018**, *77*, 180. [CrossRef]
5. Bortoloni, M.; Bottarelli, M.; Su, Y. A study on the effect of ground surface boundary conditions in modelling shallow ground heat exchangers. *Appl. Eng.* **2017**, *111*, 1371–1377. [CrossRef]
6. Gabrielli, L.; Bottarelli, M. Financial and economic analysis for ground coupled heat pump using shallow ground heat exchangers. *Sustain. Cities Soc.* **2016**, *20*, 71–80. [CrossRef]
7. Arrizabalaga, I.; De Gregorio, M.; De Santiago, C.; García de la Noceda, C.; Pérez, P.; Urchueguía, J.F. Geothermal Energy Use, Country Update for Spain. In Proceedings of the European Geothermal Congress 2019, Den Haag, The Netherlands, 11–14 June 2019.
8. Manzella, A.; Serra, D.; Cesari, G.; Bargiacchi, E.; Cei, M.; Cerutti, P.; Conti, P.; Giudetti, G.; Lupi, M.; Vaccaro, M. Geothermal Energy Use, Country Update for Italy. In Proceedings of the European Geothermal Congress 2019, Den Haag, The Netherlands, 11–14 June 2019.
9. Papachristou, M.; Arvanitis, A.; Mendrinos, D.; Dalabakis, P.P.; Karytsas, C.; Andritsos, N. Geothermal Energy Use, Country Update for Greece (2016-2019). In Proceedings of the European Geothermal Congress 2019, Den Haag, The Netherlands, 11–14 June 2019.
10. Spitler, J.; Bernier, M. Ground-source heat pump systems: The first century and beyond. *HVAC&R Res.* **2011**, *17*, 891–894.
11. Eslami-nejad, P.; Bernier, M. A preliminary assessment on the use of phase change materials around geothermal boreholes. *ASHRAE Trans.* **2013**, *119*, 312–321.
12. Lei, H.Y.; Dai, C.S. Comparative experiment of different backfill grouts for concentric ground heat exchangers. *Grc. Trans.* **2013**, *37*, 597–600.
13. Lei, H.Y.; Zhu, N. Analysis of phase change materials (PCMs) used for borehole fill materials. *Trans. Geoth Resour. Counc.* **2009**, *33*, 92–98.
14. Wang, J.L.; Zhao, J.D.; Liu, N. Numerical simulation of borehole heat transfer with phase change material as grout. *Appl. Mech. Mater.* **2014**, *577*, 44–47. [CrossRef]

15. Yang, W.; Xu, R.; Yang, B.; Yang, J. Experimental and numerical investigations on the thermal performance of a borehole ground heat exchanger with PCM backfill. *Energy* **2019**, *174*, 216–235. [CrossRef]

16. Bottarelli, M.; Bortoloni, M.; Su, Y.; Yousif, C.; Aydın, A.A.; Georgiev, A. Numerical analysis of a novel ground heat exchanger coupled with phase change materials. *Appl. Eng.* **2015**, *88*, 369–375. [CrossRef]

17. Corberán, J.M.; Cazorla-Marín, A.; Marchante-Avellaneda, J.; Montagud, C. Dual source heat pump, a high efficiency and cost-effective alternative for heating, cooling and DHW production. *Int. J. Low-Carbon Tec.* **2018**, *13*, 161–176.

18. Grossi, I.; Dongellini, M.; Piazzi, A.; Morini, G.L. Dynamic modelling and energy performance analysis of an innovative dual-source heat pump system. *Appl. Eng.* **2018**, *142*, 745–759. [CrossRef]

19. Wang, H.; Liu, X.; Feng, G.; Kang, Z.; Luo, Y.; Bai, B.; Chi, L. Simulation analysis of air-ground dual source heat pump operating efficiency. *Proced. Eng.* **2015**, *121*, 1413–1419. [CrossRef]

20. Cannistraro, M.; Mainardi, E.; Bottarelli, M. Testing a dual-source heat pump. Math Model. *Eng. Probl.* **2018**, *5*, 205–210.

21. Bottarelli, M.; Bortoloni, M.; Su, Y. On the sizing of a novel Flat-panel ground heat exchanger in coupling with a dual-source heat pump. *Renew. Energy* **2019**, *142*, 552–560. [CrossRef]

22. Phase Change Material Products Limited. Available online: http://www.pcmproducts.net/ (accessed on 20 March 2020).

23. UNI. UNI 11466:2012, Heat Pump Geothermal Systems-Design and Sizing Requirements. Available online: www.uni.com (accessed on 23 November 2018).

24. EnergyPlus Building Energy Simulation Program. Available online: https://energyplus.net/testing (accessed on 23 November 2018).

25. Bottarelli, M.; Zang, L.; Bortoloni, M.; Su, Y. Energy performance of a dual air and ground-source heat pump coupled with a Flat-panel ground heat exchanger. *Bulg. Chem. Commun.* **2016**, *48*, 64–70.

26. Nam, Y.J.; Gao, X.Y.; Yoon, S.H.; Lee, K.H. Study on the performance of a ground source heat pump system assisted by solar thermal storage. *Energies* **2015**, *8*, 13378–13394. [CrossRef]

27. Comsol Multiphysics Reference Manual. 2017. Available online: https://doc.comsol.com/5.3/doc/com.comsol.help.comsol/COMSOL_ReferenceManual.pdf (accessed on 30 March 2020).

28. Ozturk, T.; Ceber, Z.P.; Türkeş, M.; Kurnaz, M.L. Projections of climate change in the Mediterranean Basin by using downscaled global climate model outputs. *Int. J. Climatol.* **2015**, *35*, 4276–4292. [CrossRef]

29. Dongellini, M.; Naldia, C.; Morini, G.L. Annual performances of reversible air source heat pumps for space conditioning. *Energy Procedia.* **2015**, *78*, 1123–1128. [CrossRef]

Article

Experimental Verification of an Analytical Mathematical Model of a Round or Oval Tube Two-Row Car Radiator

Dawid Taler [1], Jan Taler [2],* and Marcin Trojan [1]

[1] Department of Thermal Processes, Air Protection, and Waste Utilization, Cracow University of Technology, 31-155 Cracow, Poland; dtaler@pk.edu.pl (D.T.); marcin.trojan@pk.edu.pl (M.T.)

[2] Institute of Thermal Power Engineering, Cracow University of Technology, 31-864 Cracow, Poland

* Correspondence: jtaler@pk.edu.pl

Received: 19 May 2020; Accepted: 29 June 2020; Published: 2 July 2020

Abstract: The paper presents an analytical mathematical model of a car radiator, which takes into account various heat transfer coefficients (HTCs) on each row of pipes. The air-side HTCs in a specific row of pipes in the first and second passes were calculated using equations for the Nusselt number, which were determined by CFD simulation by the ANSYS program (Version 19.1, Ansys Inc., Canonsburg, PA, USA). The liquid flow in the pipes can be laminar, transition, or turbulent. When changing the flow form from laminar to transition and from transition to turbulent, the HTC continuity is maintained. Mathematical models of two radiators were developed, one of which was made of round tubes and the other of oval tubes. The model allows for the calculation of the thermal output of every row of pipes in both passes of the heat exchangers. Small relative differences between the total heat flow transferred in the heat exchanger from hot water to cool air exist for different and uniform HTCs. However, the heat flow rate in the first row is much higher than the heat flow in the second row if the air-side HTCs are different for each row compared to a situation where the HTC is constant throughout the heat exchanger. The thermal capacities of both radiators calculated using the developed mathematical model were compared with the results of experimental studies. The plate-fin and tube heat exchanger (PFTHE) modeling procedure developed in the article does not require the use of empirical correlations to calculate HTCs on both sides of the pipes. The suggested method of calculating plate-fin and tube heat exchangers, taking into account the different air-side HTCs estimated using CFD modelling, may significantly reduce the cost of experimental research for a new design of heat exchangers implemented in manufacturing.

Keywords: tube heat exchanger with plate-fins; air-side Nusselt number; various heat transfer equations in each tube row; CFD modelling; empirical heat transfer equation

---

1. Introduction

Cross-flow plate-fin and tube heat exchangers (PFTHEs) are widely applied in industry, power plants, cars, as well as in the air conditioning and heating of buildings. If the gas temperature flowing perpendicular to the pipe axis is high, such as in evaporators, water heaters, and superheaters in heat recovery steam boilers (HRSGs) behind gas turbines, the individual round fins are welded to the outer surfaces of the pipes. If the gas temperature is slightly higher than the ambient temperature, such as in car engine coolers, air coolers in air conditioning units, evaporators and condensers in fan coils, and in so-called "dry systems" for cooling water heated in turbine condensers, aluminium tubes are usually used, on which the aluminium continuous fins are placed.

In ribbed heat exchangers with continuous fins when the air flow between the ribs is laminar, the highest HTC occurs in the first pipe row and decreases in subsequent pipe rows. This situation happens in heat exchangers with an in-line pipe array and is even more evident in heat exchangers with a staggered pipe layout. This can be explained by the very high HTC in the inlet section of the channels between the fins. On the surface of the fins, near their inlet edges, the HTC is more than ten times higher in comparison with the average coefficient over the entire surface of the fin. For continuous fins, the length of the inlet section, where fluid flow in channels formed by adjacent ribs is hydraulically and thermally developing, may be several dozen widths of the gap created by the adjacent fins.

In analytical and engineering calculation procedures such as ε-NTU (effectiveness–number of transfer units), P-NTU (effectiveness–number of transfer units), or LMTD (log mean temperature difference) method, a constant HTC on the gas side must be assumed. The ε-NTU method calculates only one effectiveness value for the whole exchanger [1], whereas the P-NTU method calculates the efficiency P separately for each medium [2]. The formulas and graphs for determining the efficiency of exchangers with typical flow systems are presented in Kuppan's book [3].

In the case of multi-pass heat exchangers with several rows of tubes, where the specific heat for one or both of these media depends on the temperature, the fluid temperature can be determined using numerical models [4]. For solving a system of partial differential equations describing the temperature of fluids, pipe walls, and fins, the finite difference method or the finite volume method may be applied.

A large number of papers deal with the determination of air-side heat transfer correlations for the estimation of the average HTC in one, two, three, and four-row PFTHEs. If a PFTHE has more than four-pipe rows, then the same air-side heat transfer correlation as for a four-row heat exchanger is used to calculate the average HTC for the entire heat exchanger. Heat transfer and friction correlations for wavy PFTHE with in-line and staggered tube arrangements were developed in [5]. Similar equations for PFTHEs with staggered pipe arrays were proposed in [6] based on experimental research. Both papers [5,6] provided relationships for the air-side Nusselt number and friction factor for single, double, and triple row PFTHEs. Performance tests of continuous plain fin and tube heat exchanger under dehumidifying conditions were reported by Wang et al. [7] and Halici et al. [8]. Studies carried out by Wang et al. [7] demonstrated that increasing the number of pipe rows causes a substantial reduction in the heat transfer characteristics of the heat transfer rate when the air is dry, and there is no condensation on the surface of the continuous fins.

The influence of the tube row number on the average HTC in the whole PFTHE was also experimentally investigated by Halici et al. [8] for PFTHEs with a staggered tube layout, which were constructed of copper tubes and aluminium fins. Four PFTHEs with numbers of rows from 1 to 4 each were studied. The friction and Colburn factors, as well as the air-side HTCs, were estimated when the air velocity varied in the range from 0.9 to 4 m/s. The experiments revealed that the friction factors and HTCs for the wet surfaces were higher than those for the dry surfaces. The average air-side HTC in the PFTHE for both wet and dry conditions deteriorated with the number of pipe rows. The HTC was approximately 21% higher for the one-row PFTHE when the air velocity was about 1 m/s. The HTC was about 14% higher compared to the four-row PFTHE for the velocity of the air equal to 3 m/s. It should be emphasized that in [5–8], correlations were determined for the Colburn parameter for the entire PFTHE and not for individual rows of tubes. Similarly, in the case of experimental investigations of a two-row car radiator presented in papers [9,10], heat transfer correlations were derived for the Nusselt number on the water and air side for the whole car radiator, and not separately for the first and second row.

Experimental studies demonstrating the influence of the pipe row number on the Colburn factor for individual tube rows and the whole heat exchanger were conducted by Rich [11]. Five PFTHEs with staggered tube arrays of five to six rows were studied. The air-side HTC decreased with row depth for a frontal air velocity calculated on a free channel cross-section lower than about 3.5 m/s. The Reynolds number evaluated for the air velocity in the narrowest free section for an equivalent hydraulic diameter equal to the tube row distance was less than about 12,000 [11].

A reliable relationship for evaluating the air pressure drop in PFTHEs with staggered tube arrays was proposed by Marković et al. [12]. Based on 872 experimental data sets, a simple equation for the friction factor of Darcy–Weisbach as a function of the Reynolds number and the ratio of the total area of the outside surface of the finned tube to the unfinned surface of the tube was developed.

The influence of other parameters, such as the shape and diameter of pipes, wall and fin thickness, and longitudinal and transverse pitch of the pipe spacing on the heat transfer in plate-fin and tube heat exchangers was discussed in books [1,3] and [13,14]. Webb and Kim [14] also studied the influence of various ways of intensifying the heat exchange on the inner surfaces of pipes to increase the thermal output of the heat exchanger.

The uniform air-side HTC on all pipe rows can also be determined by employing computer simulation using commercial CFD programs [15–17].

The paper by Sun et al. [15] deals with enhancing heat transfer in a PFTHE by using guiding channels (winglets) to direct the airflow to the back of the pipes to avoid the formation of dead zones in the region of the back stagnation point. The aim of optimizing the topology was to minimize the pressure drop on the air side in the PFTHE with constraints on the number of obstacles while improving heat exchange. The results of the simulation of the new PFTHE with the ANSYS-Fluent program were compared with experimental studies.

Performance comparison of a wavy and plain fin with radiantly distributed winglets around each tube in a PFTHE was carried out both numerically and experimentally by Li et al. [16]. The air velocity before the PFTHE varied from 1.5 to 7.5 m/s. If the velocity of the air in front of the PFTHE was between 1.5 m/s and 3.5 m/s, the laminar flow was adopted in CFD modelling and if higher than 3.5 m/s, the flow was simulated as turbulent. Equations for the air-side Nusselt number for different types of continuous fins were found in experimental studies.

Direct numerical simulation (DNS) was also recently used to estimate air-side correlations in PFTHEs. Nagaosa [17] provided the air-side heat transfer correlation based on the DNS simulation, which is in good agreement with the experimentally determined correlation. However, the disadvantage of the DNS is the very long time of computer calculations.

For large values of the air-side Reynolds numbers when the flow is turbulent, the first row Nusselt number is smaller than the Nusselt numbers on the second and further rows of pipes. An increase in the average Nusselt number on each subsequent pipe row for turbulent air flow occurs in heat exchangers made of plain [1–3] or individually finned pipes [18] and also in exchangers with continuous fins [11]. Kearney and Jacobi [18] applied laser triangulation to naphthalene sublimation experiments to determine row-by-row Nusselt numbers in cross-flow ribbed tube heat exchangers with two rows of tubes. The equations for the average Nusselt number in the first, second, and whole heat exchanger were obtained for the in-line and staggered two-row arrays in a Reynolds number range from 5000 to 28,000. The Reynolds number was calculated for the hydraulic diameter and maximum air velocity in the narrowest cross-section of the heat exchanger. For the in-line layout, the first-row Nusselt number was 34% lower than the Nusselt number in the second row for a Reynolds number of about 5000 and 45% lower for a Reynolds number of about 28,000. The first row of pipes in the exchanger with staggered pipe arrangement showed a 45% lower Nusselt number in the whole range of Reynolds numbers compared to the second row.

The opposite is true for cross-flow tube heat exchangers in steam and water boilers, such as water heaters, convection evaporators, and superheaters. In these heat exchangers, the largest HTC occurs in the first row of pipes. In front of the heat exchangers in boilers, there are large spaces containing radiating gases such as carbon dioxide or water steam. In addition to the convection heat exchange in the first row of pipes, there is also very strong irradiation of three-atomic gases in front of the heat exchanger. At high temperatures in the range from about 600 °C to 1100 °C, there is strong irradiation of the pipe circumference on the inflow side from the flue gas. The sum of convection and radiation HTC is the largest in the first row and decreases in the next rows of pipes.

Particularly high heat fluxes are taken up by ribbed tubes situated in the first row of tubes, which are usually used in HRSGs and gas-fired water or steam boilers. For this reason, the pipes in the first row are subject to premature wear due to the overheating of the pipe wall and fins on the inlet side.

The analysis of the works published so far shows that there are no mathematical models of heat exchangers taking into account different HTCs in particular rows of pipes. This paper develops an exact analytical mathematical model of a two-pass car radiator, taking into account the different air-side HTCs in the first and second row of tubes. The results of modelling car radiators made of round and oval pipes were compared with the results of experimental research. The water flow in the pipes can be laminar, transition or turbulent while maintaining the continuity of the heat transfer coefficient when changing the flow regime. A calculation method of PFTHs was proposed without using empirical heat transfer correlations on the water and air sides. CFD modelling does not have such an advantage at this stage of development.

2. Mathematical Model of the Two-Pass PFTHE accounting for Different HTCs in Every Tube Row

An analytical mathematical model of an automobile radiator with two rows of pipes was developed. The HTCs in the first and second tube row were calculated using heat transfer correlations, which were found based on CFD (computational fluid dynamics) simulations. The flow diagram of the coolers analysed in the study is depicted in Figure 1.

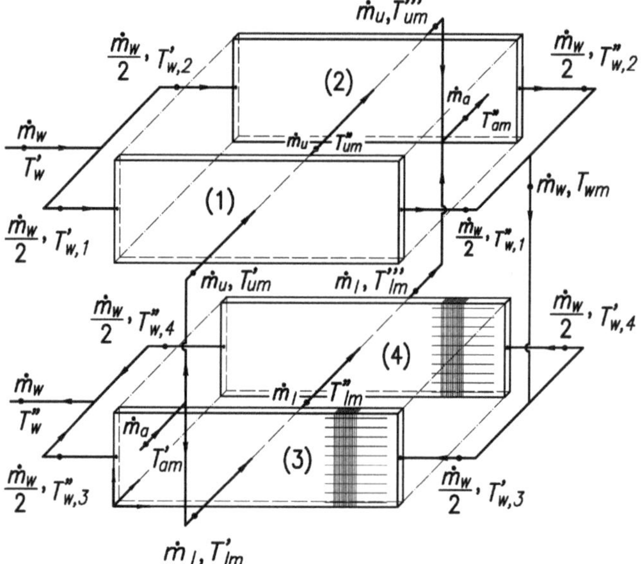

Figure 1. Flow arrangement of the car radiator analysed in the study: the two-pass car radiator with two rows of tubes consists of the first tube row in the first pass (1), second tube row in the first pass (2), first tube row in the second pass (3), and second tube row in the second pass (4).

The flow rate of the water $\dot{m}_w$ to the engine cooler is divided into two equal parts $\dot{m}_w/2$ flowing through both rows of pipes in the first pass. The starting point for building a mathematical model of an engine cooler is a single-pass of the double-row PFTHE (first pass in the two-pass PFTHE with two tube rows) (Figure 2).

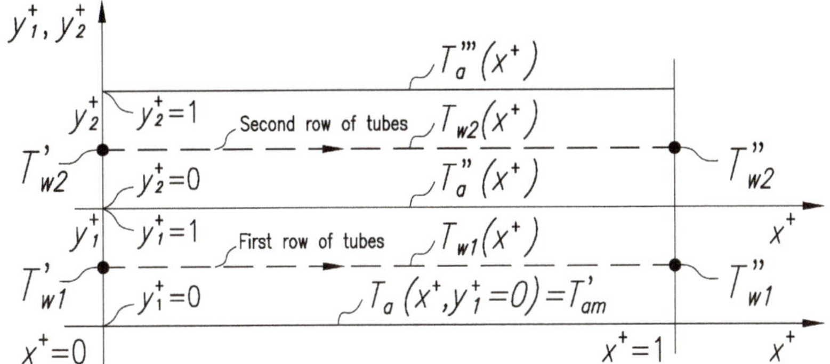

Figure 2. Two rows of tubes in a single-pass heat exchanger.

2.1. Analytical Model for the First Row of Pipes

The governing energy conservation equations for the water and air for the tube situated in the first row are

$$\frac{d\,T_{w,\,1}\left(x^+\right)}{d\,x^+} = -N_w^I\left[\,T_{w,\,1}\left(x^+\right) - T_{m\,a}^I\left(x^+\right)\right]\quad 0 \le x^+ \le 1 \tag{1}$$

$$\frac{\partial\,T_a^I\left(x^+,y_1^+\right)}{\partial\,y_1^+} = N_a^I\left[\,T_{w,\,1}\left(x^+\right) - T_a^I\left(x^+,y_1^+\right)\right]\quad 0 \le x^+ \le 1\;\; 0 \le y_1^+ \le 1 \tag{2}$$

where $x^+ = x/L_c$, $y_1^+ = y_1/p_2$ dimensionless coordinates x, y_1 Cartesian coordinates (Figure 2) $T_{w,1}$ the water temperature in the tube in the first row, T_a^I the temperature of the air in the first row of pipes.

Two local dimensionless coordinate systems were introduced $\left(x^+,y_1^+\right)$ and $\left(x^+,y_2^+\right)$ (Figure 2). The following expression determines the average air temperature over the first-row width

$$T_{m\,a}^I\left(x^+\right) = \int_{y_1^+=0}^{1} T_a^I\left(x^+,y_1^+\right) dy_1^+ \tag{3}$$

The number of heat transfer units (NTU) on the water-side N_w^I and air-side N_a^I is defined as follows

$$N_w^I = 2U^I\,A_{bout}^I/\left(\dot{m}_w\,\bar{c}_{p\,w}\right)\quad N_a^I = U^I\,A_{bout}^I/\left(\dot{m}_a\,\bar{c}_{p\,a}\right) \tag{4}$$

where $A_{bout}^I = n_u^I P_{out} L_c$, n_u^I the total number of pipes in the first row of the first pass, A_{bout}^I outer surface area of the bare pipes in the first row of the first pass, P_{out} the outer circumference of the plain tube, L_c length of a tube in the PFTHE. The symbol $\bar{c}_{pw}$ is the average specific heat of the water in the temperature range from T_w'' to T_w' and the average specific heat of the air $\bar{c}_{pa}$ in the temperature range from T_{am}' to T_{am}'' (Figure 2).

The overall HTC U^I is calculated using the expression

$$\frac{1}{U^I} = \frac{A_{bout}^I}{A_{bin}^I}\frac{1}{h_{in}^I} + \frac{A_{bout}^I}{A_{wm}^I}\frac{\delta_w}{k_w} + \frac{1}{h_{oe}^I} \tag{5}$$

where $A_{bin}^I = n_u^I P_{in} L_c$, the inner surface area of the pipes in the first row of the first pass, P_{in} the inner circumference of the tube, h_{in}^I HTC at the inner surface of the pipe, $A_{wm}^I = \left(A_{bout}^I + A_{bin}^I\right)/2$ mean surface area of the plain pipe in the first row, δ_w the thickness of the tube wall, k_w the thermal conductivity of the tube material.

The equivalent HTC h_{oe}^I, considering the heat exchange through the fins fixed to the outer surface of a plain tube, is given by

$$h_{oe}^I = h_a^I \left[\frac{A_{bf}^I}{A_{bout}^I} + \frac{A_f^I}{A_{bout}^I} \eta_f\left(h_a^I\right) \right] \tag{6}$$

where h_a^I air-side HTC in the first row of tubes in the first pass, A_f^I fin surface area, A_{bf}^I tube surface area between fins, η_f fin efficiency.

The boundary conditions for differential Equations (1) and (2) are

$$T_{w,1}\big|_{x^+ = 0} = T'_{w,1} \tag{7}$$

$$T_a^I\big|_{y_1^+ = 0} = T'_{am} \tag{8}$$

The solution to Equations (1) and (2) with boundary conditions (7) and (8) is

$$T_{w,1}\left(x^+\right) = T'_{am} + \left(T'_{w,1} - T'_{am}\right) \exp\left\{-\frac{N_w^I}{N_a^I}\left[1 - \exp\left(-N_a^I\right)\right]x^+\right\} \quad 0 \le x^+ \le 1 \tag{9}$$

$$T_a^I\left(x^+, y_1^+\right) = T_{w,1}\left(x^+\right) - \left[T_{w,1}\left(x^+\right) - T'_{am}\right] \exp\left(-N_a^I y_1^+\right) \quad 0 \le x^+ \le 1 \; 0 \le y_1^+ \le 1 \tag{10}$$

The air temperature $T_a''(x^+)$ after the first row of pipes is obtained from Equation (10) after substitution $y_1^+ = 1$

$$T_a''\left(x^+\right) = T_a^I\left(x^+, y_1^+ = 1\right) = T_{w,1}\left(x^+\right) - \left[T_{w,1}\left(x^+\right) - T'_{am}\right] \exp\left(-N_a^I\right) \tag{11}$$

The temperature $T_{w,1}(x^+)$ in Equation (11) is defined by Equation (9). The mean air temperature behind the first row of pipes T_{am}'' is obtained from its definition

$$T_{am}'' = \int_0^1 T_a''\left(x^+\right) dx = T'_{am} + \frac{N_a^I}{N_w^I}\left(T'_{w,1} - T'_{am}\right)\left[1 - \exp(-B)\right] \tag{12}$$

where B is as follows:

$$B = \frac{N_w^I}{N_a^I}\left[1 - \exp\left(-N_a^I\right)\right] \tag{13}$$

The temperature distribution of water and air in the first tube row can be determined using Equations (9)–(13).

2.2. Analytical Model for the Second Row of Pipes

The relevant differential equations for the second row of pipes are

$$\frac{d\,T_{w,2}\left(x^+\right)}{d\,x^+} = -N_w^{II}\left[T_{w,2}\left(x^+\right) - T_{m\,a}^{II}\left(x^+\right)\right] \quad 0 \le x^+ \le 1 \tag{14}$$

$$\frac{\partial\,T_a^{II}\left(x^+, y_2^+\right)}{\partial\,y_2^+} = N_a^{II}\left[T_{w,2}\left(x^+\right) - T_a^{II}\left(x^+, y_2^+\right)\right] \quad 0 \le x^+ \le 1 \; 0 \le y_2^+ \le 1 \tag{15}$$

where

$$N_w^{II} = 2U^{II} A_{bout}^{II}/\left(\dot{m}_w \bar{c}_{p\,w}\right) \quad N_a^{II} = U^{II} A_{bout}^{II}/\left(\dot{m}_a \bar{c}_{p\,a}\right) \tag{16}$$

The overall HTC U^{II} can be calculated using Equation (5), in which the HTC h_{oe}^I is to be substituted by h_{oe}^{II}. Expression (6) is used to calculate the effective air-side HTC h_{oe}^{II}, but instead of the HTC, h_a^I, the HTC h_a^{II} for the second row of pipes must be substituted.

The average air temperature across the width of the second tube row $T^{II}_{m\,a}(x^+)$ is determined as follows

$$T^{II}_{m\,a}\left(x^+\right) = \int\limits_{y^+_2=0}^{1} T^{II}_a\left(x^+, y^+_2\right) dy^+_2 \tag{17}$$

The water temperature at the inlet to the second row of pipes is $T'_{w,\,2}$, and the air temperature in front of the second row of tubes is equal to the outlet temperature $T''_a(x^+)$ of the air from the first row of pipes. Thus, the boundary conditions are as follows:

$$T_{w,\,2}\big|_{x^+=0} = T'_{w,\,2} \tag{18}$$

$$T^{II}_a\big|_{y^+_2=0} = T''_a\left(x^+\right) \tag{19}$$

The solution to Equations (14) and (15) under boundary conditions (18) and (19) is

$$T_{w,\,2}(x^+) = T'_{am} + \tfrac{E}{D-B}\left[\exp(-B\,x^+) - \exp(-D\,x^+)\right] + \left(T'_{w,\,2} - T'_{am}\right)\exp(-D\,x^+)\quad 0 \le x^+ \le 1 \tag{20}$$

$$T^{II}_a\left(x^+, y^+_2\right) = T_{w,\,2}\left(x^+\right) - \left[T_{w,\,2}\left(x^+\right) - T''_a\left(x^+\right)\right]\exp\left(-N^{II}_a\,y^+_2\right) 0 \le x^+ \le 1\ 0 \le y^+_2 \le 1 \tag{21}$$

where

$$D = \frac{N^{II}_w}{N^{II}_a}\left[1 - \exp\left(-N^{II}_a\right)\right]\quad E = D\left(T'_{w,\,1} - T'_{am}\right)\left[1 - \exp\left(-N^I_a\right)\right] \tag{22}$$

The coefficient B is given by Equation (13). The expression for air temperature after the second row of pipes $T'''_a(x^+)$ (Figure 2) is obtained from Equation (21), taking into account that $y^+_2 = 1$. After substitution of Equation (20) into Equation (21) and transformations, the following expression for the air temperature $T'''_a(x^+)$ behind the second tube row is obtained

$$\begin{aligned}
T'''_a\left(x^+\right) = T^{II}_a\left(x^+,\ y^+_2 = 1\right) = T'_{am} &+ \left[1 - \exp\left(-N^{II}_a\right)\right]\left\{\tfrac{E}{D-B}\left[\exp(-B\,x^+) - \exp(-D\,x^+)\right] + \right.\\
&\left. + \left(T'_{w,\,2} - T'_{am}\right)\exp(-D\,x^+)\right\} + \left(T'_{w,\,1} - T'_{am}\right)\left[1 - \exp\left(-N^I_a\right)\right]\exp\left(-B\,x^+ - N^{II}_a\right) \\
&0 \le x^+ \le 1
\end{aligned} \tag{23}$$

The mean air temperature T'''_{am} leaving the second tube row T'''_{am} is calculated as follows

$$\begin{aligned}
T'''_{am} = \int\limits_0^1 T'''_a\left(x^+\right) dx^+ = T'_{am} &+ \left[1 - \exp\left(-N^{II}_a\right)\right]\left\{\tfrac{E}{(D-B)\,B}\left[1 - \exp(-B)\right] - \right.\\
&\left. - \tfrac{E}{(D-B)\,D}\left[1 - \exp(-D)\right] + \tfrac{\left(T'_{w,\,2} - T'_{am}\right)}{D}\left[1 - \exp(-D)\right]\right\} - \\
&- \tfrac{\left(T'_{w,\,1} - T'_{am}\right)}{B}\left[1 - \exp\left(-N^I_a\right)\right]\left[\exp\left(-B - N^{II}_a\right) - \exp\left(-N^{II}_a\right)\right]
\end{aligned} \tag{24}$$

When the air-side HTC h^I_a in the first pipe row is equal to the HTC h^{II}_a in the second pipe row, then $D = B$ and the denominator $(D-B)$ is zero. When $D = B$, Equation (20) requires transformation. Writing Equation (20) as

$$T_{w,\,2}\left(x^+\right) = T'_{am} + \frac{E\,\exp(-D\,x^+)}{D - B}\left\{\exp\left[(D - B)\,x^+\right] - 1\right\} + \left(T'_{w,2} - T'_{am}\right)\exp\left(-D\,x^+\right) \tag{25}$$

and applying the L'Hôpital rule to an expression where, at $D = B$, the numerator and the denominator are equal to zero, one obtains

$$\lim_{D\to B}\frac{\exp[(D-B)\,x^+]-1}{D-B} = \lim_{z\to 0}\frac{\exp(z\,x^+)-1}{z} = \lim_{z\to 0}\frac{\frac{d}{dz}[\exp(z\,x^+)-1]}{\frac{dz}{dz}} = \lim_{z\to 0}\frac{x^+\,\exp(z\,x^+)}{1} = x^+ \tag{26}$$

where $z = D - B$.

Taking into account the relation (26) for $D = B$ (i.e., for $h_a^I = h_a^{II} = h_a$) in Equation (25) gives

$$T_{w,2}(x^+) = T'_{am} + \left\{ B\left(T'_{w,1} - T'_{am}\right)\left[1 - \exp\left(-N_a^I\right)\right] x^+ + \left(T'_{w,2} - T'_{am}\right)\right\} \exp(-B\,x^+) \quad 0 \le x^+ \le 1 \quad (27)$$

where $N_a^I = N_a^{II} = N_a$.

If $h_a^I = h_a^{II} = h_a$, then the air temperature $T_a'''\,(x^+)$ after the second row of pipes is obtained with Equation (23) after taking into account the relation (26)

$$T_a'''\,(x^+) = T_a^{II}\left(x^+,\, y_2^+ = 1\right) = T'_{am} + \left[1 - \exp\left(-N_a^I\right)\right] \times$$
$$\times\left\{\left[C\,x^+ + \left(T'_{w,1} - T'_{am}\right)\right]\exp(-B\,x^+) + \left(T'_{w,1} - T'_{am}\right)\exp\left[-\left(B\,x^+ + N_a^I\right)\right]\right\} \quad 0 \le x^+ \le 1 \quad (28)$$

where the symbol C is as follows:

$$C = B\left(T'_{w,1} - T'_{am}\right)\left[1 - \exp\left(-N_a^I\right)\right] \quad (29)$$

3. An Analytical Model of an Internal Combustion Engine Radiator

The equations derived for the single-pass double-row PFTHE, taking into account the different HTCs in individual tube rows, were used to develop an analytical model of the entire radiator depicted in Figure 1. This radiator is used in spark engines with a displacement capacity of 1600 cubic centimeters.

A description of the radiator construction made of round tubes is presented in [10], and the radiator built of oval tubes in [9]. First, the temperatures $T''_{w,1}$ and $T''_{w,2}$ of the liquid at the outlet from the first pass of the cooler are calculated. Water from the first and second row of pipes is mixed in the reversing manifold connecting the outlet from the first pass with the inlet to the second pass.

The water mixture with temperature T_{wm} (Figure 1) feeds the second pass of the automobile radiator. The water mass flow rate $\dot{m}_w$ is the same as in the first pass. The stream of the water is divided into two equal parts $\dot{m}_w/2$ flowing through the first and second tube rows. The air flows evenly across the entire radiator cross-section. The rate of the air mass flow across the first (upper) pass is $\dot{m}_a n_u / (n_u + n_l)$ and that of the second (lower) pass is $\dot{m}_a n_l / (n_u + n_l)$. The symbol $\dot{m}_a$ is the mass flow rate of air in front of the radiator. The symbols n_u and n_l are the respective numbers of pipes in the first and second passes in the first row of tubes. The numbers n_u and n_l are equal to ten and nine, respectively. Thermal calculations of the second (lower) pass were carried out analogously to those of the first (upper) pass. The comparison of characteristics of oval and round tube are shown in Table 1.

Table 1. Characteristics of finned tubes in the heat exchanger from oval and round tubes.

Geometric Data	The Heat Exchanger of Oval Pipes	The Heat Exchanger of Round Pipes
Outer tube diameter d_o, mm	Maximum tube diameter $d_{o,max} = 11.82$ mm Minimum tube diameter $d_{o,min} = 6.35$ mm	$d_o = 7.2$ mm
Tube wall thickness δ_w, mm	$\delta_w = 0.4$ mm	$\delta_w = 0.5$ mm
Height p_1 and width p_2 of the apparent fin associated with one tube, mm	$p_1 = 18.5$ mm, $p_2 = 17$ mm	$p_1 = 18.5$ mm, $p_2 = 12$ mm
Fin thickness δ_f, mm	$\delta_f = 0.08$ mm	$\delta_f = 0.08$ mm
Air – side hydraulic diameter d_{ha}, mm	$d_{ha} = 1.41$ mm	$d_{ha} = 1.95$ mm
Water – side hydraulic diameter d_{hw}, mm	$d_{hw} = 7.06$ mm	$d_{hw} = d_o - 2\,\delta_w = 6.2$ mm

The thermal conductivity of aluminium tubes and fins was assumed to be $k_f = k_w = 207$ W/(m·K).

4. The Air-Side Heat Transfer Correlations

The air-side HTCs were estimated by CFD simulation using ANSYS CFX. The method of determining the correlations for the average Nusselt number in the individual row of tubes and the whole radiator was presented in [19].

Nusselt number correlations for PFTHE from oval and round pipes are given below:

- oval pipe radiator

$$Nu_a^I = 30.7105 Re_a^{-0.24} Pr_a^{1/3} \quad 150 \leq Re_a \leq 330 \tag{30}$$

$$Nu_a^{II} = 0.0744 Re_a^{0.7069} Pr_a^{1/3} \quad 150 \leq Re_a \leq 330 \tag{31}$$

$$Nu_a = 1.0605 Re_a^{0.2974} Pr_a^{1/3} \quad 150 \leq Re_a \leq 330 \tag{32}$$

The Nusselt number is defined as $Nu_a = h_a d_{ha}/k_a$, where k_a is the thermal conductivity of the air. The Reynolds number Re_a on the air side is based on the hydraulic diameter and air velocity at the minimum flow area. The hydraulic diameter for the analysed oval pipe cooler was $d_{h\,a} = 1.42$ mm. The exponent of the Reynolds number is very small, which indicates that the average Nusselt number in the first row is almost constant. A constant Nusselt number is a characteristic feature of hydraulically and thermally developed laminar flow. The effect of vortices forming near the forward and backward stagnation point is more pronounced in the second tube row, as the power exponent of the Reynolds number is larger than in the first row. The heat transfer correlation (32) for the average Nusselt number in a two-row bundle is similar to the expression for the average Nusselt number on a flat surface for a thermally and hydraulically developing laminar flow, for which the power exponent of the Reynolds number is 1/3. The following are heat transfer correlations for calculating the Nusselt number on the air side for an engine radiator manufactured from circular tubes.

- round pipe radiator

$$Nu_a^I = 1.6502 Re_a^{0.2414} Pr_a^{1/3} \quad 100 \leq Re_a \leq 525 \tag{33}$$

$$Nu_a^{II} = 0.1569 Re_a^{0.5499} Pr_a^{1/3} \quad 100 \leq Re_a \leq 525 \tag{34}$$

$$Nu_a = 0.6070 Re_a^{0.3678} Pr_a^{1/3} \quad 100 \leq Re_a \leq 525 \tag{35}$$

A comparison of correlations (33) and (34) demonstrates that laminar flow predominates in the first row of round tubes, as the exponent of the Reynolds number is low and equals 0.2414. In the second row of pipes, there are vortexes at the front and back of the pipe, and the exponent of the Reynolds number is higher and amounts to 0.5499. However, the average Nusselt number in the first row is higher than that in the second row, which is mainly due to the inlet section in the gap formed by the two adjacent fins, characterized by very high heat transfer coefficients on the fin surfaces. As a result, the heat flow rate transferred to the air in the first row is higher than that in the second row. The reduction of the heat transfer from the pipe surface and the fins in the second pipe row is strongly influenced by the vortexes forming near the front and backward stagnation point on the pipe surface. The rotating air has a temperature close to that of the fin and pipe surface, and from the point of view of heat exchange, these are dead zones.

The comparison of heat flow rates obtained by the method developed with the experimental results was conducted for two various car radiators. The flow system of both radiators is the same (Figure 1). The first radiator is manufactured from circular pipes and the second one from oval pipes.

5. The Liquid-Side Heat Transfer Correlations

One of the objectives of this study is to illustrate that by the use of theoretical relationships to calculate the HTC on the inner surfaces of the heat exchanger pipes and the use of the liquid-side heat transfer correlations determined by CFD modelling, very similar results can be obtained as with experimental HTCs. By using known and experimentally tested correlations to calculate HCTs, experimental research costs can be significantly reduced or even eliminated entirely. Therefore, Gnielinski's formulas presented in the VDI (Verein Deutscher Ingenieure) Heat Atlas [20] were applied to determine the HTC in the laminar flow and relationships proposed by Taler [21] for transition and turbulent flow in tubes. In the range of the laminar flow regime, the average HTC along the entire length of pipe L_t, assuming that the velocity distribution at the cross-section of the pipe inlet is flat and the liquid flow is hydraulically and thermally developing, is determined by the following formula [20]

$$\mathrm{Nu}_{m,\,q} = \left[\mathrm{Nu}^3_{m,\,q,\,1} + 0.6^3 + \left(\mathrm{Nu}_{m,\,q,\,2} - 0.6\right)^3 + \mathrm{Nu}^3_{m,\,q,\,3}\right]^{1/3} \quad \mathrm{Re}_w \leq 2\,300 \tag{36}$$

The symbol $\mathrm{Nu}_{m,q,1}$ designates the average Nusselt number in the developed laminar flow

$$\mathrm{Nu}_{m,\,q,\,1} = \frac{48}{11} = 4.364 \quad \mathrm{Re}_w \leq 2\,300 \tag{37}$$

The second Nusselt number $\mathrm{Nu}_{m,\,q,\,2}$ represents the Lévêque solution [4] for developing flow over the planar surface, and $\mathrm{Nu}_{m,\,q,\,3}$ was derived numerically for the constant liquid velocity at the tube inlet

$$\mathrm{Nu}_{m,\,q,\,2} = 3^{1/3}\Gamma(2/3)\left(\mathrm{Re}_w\mathrm{Pr}_w\frac{d_{hw}}{L_t}\right)^{1/3} = 1.9530\left(\mathrm{Re}_w\mathrm{Pr}_w\frac{d_{hw}}{L_t}\right)^{1/3} \quad \mathrm{Re}_w \leq 2\,300 \tag{38}$$

$$\mathrm{Nu}_{m,\,q,\,3} = 0.924\left(\mathrm{Re}_w\frac{d_{hw}}{L_t}\right)^{1/2}\mathrm{Pr}_w^{1/3} \quad \mathrm{Re}_w \leq 2\,300 \tag{39}$$

The Nusselt number in the transition and turbulent flow range was evaluated by the Taler correlation [21]:

$$\mathrm{Nu}_w = \left.\mathrm{Nu}_{m,q}\right|_{\mathrm{Re}_w=2\,300} + \frac{\frac{\xi_w}{8}(\mathrm{Re}_w - 2\,300)\,\mathrm{Pr}_w^{1.008}}{1.084 + 12.4\sqrt{\frac{\xi_w}{8}}\left(\mathrm{Pr}_w^{2/3} - 1\right)}\left[1 + \left(\frac{d_{hw}}{L_t}\right)^{2/3}\right] \quad 2300 \leq \mathrm{Re}_w \leq 10^6$$
$$0.1 \leq \mathrm{Pr}_w \leq 1\,000,\ \frac{d_{hw}}{L_t} \leq 1 \tag{40}$$

The Darcy–Weisbach friction factor for turbulent flow, when $3000 \leq \mathrm{Re}_w$ was calculated by the Taler equation [22]:

$$\xi_w = (1.2776\,\log\mathrm{Re}_w - 0.406)^{-2.246} \quad 3000 \leq \mathrm{Re}_w \tag{41}$$

The relationship for the coefficient of friction for the transition region was obtained by the linear interpolation [22] between the value $\xi_w = 64/2300 = 0.02783$ for $\mathrm{Re}_w = 2300$ and $\xi_w(\mathrm{Re}_w = 3000) = 0.04355$

$$\xi_w = 0.02783 + 2.2457 \cdot 10^{-5}(\mathrm{Re}_w - 2\,300) \quad 2\,300 \leq \mathrm{Re}_w \leq 3\,000 \tag{42}$$

The Reynolds number on the water side $\mathrm{Re}_w = w_w\,d_{h\,w}/\nu_w$ was determined using hydraulic diameter d_{hw}. The hydraulic diameter was $d_{hw} = 7.06$ mm for oval tubes and $d_{hw} = d_{in} = 6.2$ mm for circular tubes. The water's physical properties were evaluated at the average temperature $\overline{T}_w = (T'_w + T''_w)/2$.

6. Experimental Results

Using the results of experimental research on two car radiators, one of which is made of oval tubes and the other of round tubes, heat transfer correlations on the air side were found. The HTCs on the inner surface of pipes were calculated using Equations (36)–(42), depending on the laminar, transition, or turbulent flow mode.

6.1. Engine Cooler Made of Oval Tubes

Experimental research was conducted at a test facility presented in [10]. The unknown parameters, the coefficient, and power exponent of the Reynolds number Re_a were determined based on 48 measurement series. The velocity of the air before the radiator varied between 0.98 and 2.01 m/s, and the water volume flow rate at the radiator inlet was between 627 and 2421 litres/h. The air temperature in front of the PFTHE ranged from 11.0 to 15.0 °C, and the temperature of the water at the car radiator inlet was between 59.6 and 72.2 °C. The air side Reynolds number ranged from 150 to 330, and the water-side Reynolds number in the first (upper) pass varied from 2500 to 11,850. When determining the correlations on the air-side, the water-side HTC was calculated using Equations (36)–(42). The following Equation for the Nusselt number Nu_a on the air side was found using the experimental data

$$Nu_a = x_1 Re_a^{x_2} Pr_a^{1/3} \quad 150 \leq Re_a \leq 330 \tag{43}$$

where the parameters x_1 and x_2 are

$$x_1 = 0.5162 \pm 0.0116 \quad x_2 = 0.4100 \pm 0.0041 \tag{44}$$

Numbers with a sign ± in Equation (44) represent half of the 95% confidence interval.

6.2. Engine Cooler Made of Round Tubes

Flow and heat measurements of a car radiator made of round pipes were carried out on the stand presented in paper [10]. The air-side heat transfer correlation was determined using 70 measurement sets. In the measurement tests, the air velocity, flow rate, and temperature of the water at the car radiator inlet were changed.

The velocity of the air changed between 0.74 and 2.27 m/s, and the water flow rate at the radiator inlet was between 300 and 2410 litres/h. The air temperature before the car radiator ranged from 3.6 to 15.5 °C, and the temperature of the water at the inlet to the radiator was between 50 and 72 °C. The water flow regime changed from laminar to transitional and then to turbulent during testing. The air-side Reynolds Re_a number ranged from 150 to 560. The Reynolds number $Re_{w,u}$ on the water side in the first pass of the heat exchanger varied from 1450 to 16,100 during the test. The water-side HTC in round tubes was evaluated using relationships (36)–(42). The following Equation for the Nusselt number Nu_a on the air side was obtained by the least-squares method

$$Nu_a = x_1 Re_a^{x_2} Pr_a^{1/3} \quad 225 \leq Re_a \leq 560 \tag{45}$$

Based on the 70 data series, the following values of parameters x_1 and x_2 were found

$$x_1 = 0.5248 \pm 0.1572 \quad x_2 = 0.4189 \pm 0.0501 \tag{46}$$

Numbers with a sign ± in Equation (46) represent half of the 95% confidence interval.

7. Analysis and Discussion of Heat Flow Rates from Water to Air Transferred in the Entire Heat Exchanger and Individual Rows of Pipes

For a radiator constructed from oval pipes, the results of mathematical modelling and measurements are presented versus the Reynolds number Re_a on the air side for two selected volume flow rates of water $\dot{\overline{V}}_w = 326.1$ litres/h and $\overline{V}_w = 1273.4$ litres/h. For a car radiator built of round tubes, the results of tests and calculations are presented versus the water Reynolds number $Re_{w,u}$ for the first pass of the heat exchanger for the preset velocity of the air w_0 before the radiator equal to 2.27 m/s.

7.1. Engine Cooler Made of Oval Tubes

The thermal capacity of the radiator was calculated by using the radiator analytical model developed in this study, assuming the uniform air-side HTC on both rows of pipes. Figure 3a shows heat flow rates estimated using the analytical model of the PFTHE for mean volume flow rate $\overline{\dot{V}}_w = 326.06$ liters/h, average air temperature at the inlet $\overline{T}'_{am} = 13.62\ °C$, and average water inlet temperature $\overline{T}'_w = 59.61\ °C$. The mean values $\overline{\dot{V}}_w$, $\overline{T}'_{am}$, and $\overline{T}'_w$ were calculated for seven measured data series. The water flow in the tubes was laminar. The water side Reynolds number Re_w in the first (upper) pass ranged from 1,222 to 1287, and in the second pass from 1358 to 1430. The air velocity w_0 varied between 0.71 m/s and 2.2 m/s. Figure 3b presents heat flow rates obtained for the following data: $\overline{\dot{V}}_w = 1273.37$ liters/h, $\overline{T}'_{am} = 14.28\ °C$, and $\overline{T}'_w = 60.51\ °C$. The water flow in the pipes was turbulent as the Reynolds number in the first pass of the cooler varied from 5238 to 5440, and in the second pass ranged from 5820 to 5883. The air velocity in front of the heat exchanger changed from 0.71 m/s to 2.2 m/s.

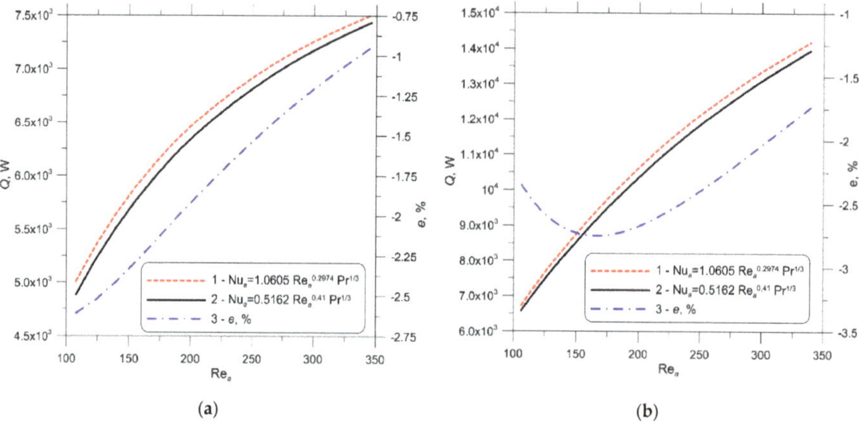

Figure 3. The heat transfer rate $Q_{t,1HTC}$ from water to air in an engine radiator made of oval tubes with an equal HTC h_a on the air side of the entire radiator; (a) $\overline{\dot{V}}_w = 326.06$ litres/h, $\overline{T}'_{am} = 13.62\ °C$ and $\overline{T}'_w = 59.61\ °C$; (b) $\overline{\dot{V}}_w = 1273.37$ litres/h, $\overline{T}'_{am} = 14.28\ °C$ and $\overline{T}'_w = 60.51\ °C$; 1—air side equation for Nu_a obtained from the CFD modelling, 2—empirical air side equation for Nu_a, 3–relative difference e.

The HTC on the water side was calculated with a mathematical model of a radiator using Equations (36)–(42). The air-side HTC was calculated using Equation (32) based on CFD simulation (Equation (1) in Figure 3) or empirical Equation (43) (Equation (2) in Figure 3).

The inspection of the results depicted in Figure 3a,b illustrates that the absolute value of the difference e does not exceed 2.75%. The heat flow rates based on the measurement data $Q_{w,\ exp}$ and CFD simulations $Q_{w,\ calc}$ were calculated as follows

$$Q_{w,\ exp} = \dot{V}_w\ \rho_w\!\left(T'_{w,\ exp}\right) \overline{c}_w\!\left(T'_{w,\ exp} - T''_{w,\ exp}\right) \tag{47}$$

$$Q_{w,\ calc} = \dot{V}_w\ \rho_w\!\left(T'_{w,\ exp}\right) \overline{c}_w\!\left(T'_{w,\ exp} - T''_{w,\ calc}\right) \tag{48}$$

where $\overline{c}_w$-the average specific heat of the water in the interval between the outlet $T''_{w,exp}$ (Equation (47)) or $T''_{w,calc}$ (Equation (48)) and inlet water temperature $T'_{w,exp}$, $\dot{V}_w$-volumetric water flow rate measured at the car radiator inlet.

The relative difference e was obtained from the following Equation

$$e = \frac{Q_{w,\,exp} - Q_{w,\,calc}}{Q_{w,\,exp}} 100, \; \% \tag{49}$$

Figure 4 shows the measurement and calculation results for the heat transfer rate Q for the seven measurement series and the calculation results for the averaged measurement data: $\dot{V}_w = 326.06$ litres/h, $\overline{T}'_{am} = 13.62$ °C, and $\overline{T}'_w = 59.61$ °C. The heat flow rate values Q (Entries 2 and 3) were determined using an analytical model of the engine cooler, in which the water-side HTC was calculated using Equations (36)–(42). Because of the small water flow rate, the flow regime of water in the heat exchanger pipes was laminar. The HTC on the air side was determined either by Equation (32) obtained from CFD simulation (Figure 4a) or by empirical correlation (43) (Figure 4b). The values of the relative difference e, calculated using Equation (49), are also shown in Figure 4. The analysis of the results shown in Figure 4a,b indicates that the use of correlation (32) based on CFD modelling for the calculation of the air-side HTC gives an excellent agreement between the calculated and measured heat flow rates. When using both CFD-based Equation (32) or empirical Equation (43), the relative differences e range from about (-1.5%) to about 5.2%.

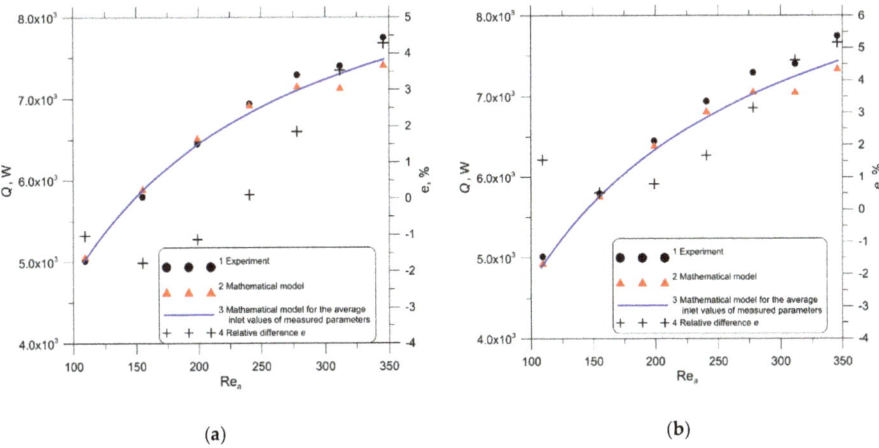

(a)
(b)

Figure 4. Heat flow rate Q transferred from hot water to cool air in the car radiator; 1: $Q_{w,\,exp}$ values determined using the measurement results, 2: heat flow rate $Q_{w,\,calc}$ obtained for the air-side Nusselt number Nu_a estimated by CFD simulation (**a**) or empirical correlation (**b**), 3: heat flow rate $Q_{w,\,calc}$ calculated using the CFD-based Equation (**a**) or empirical Equation (**b**) and the average input data from seven measurement series: $\dot{V}_w = 326.06$ liters/h, $\overline{T}'_{am} = 13.62$ °C and $\overline{T}'_w = 59.61$ °C, 4: the relative difference between $Q_{w,\,exp}$ and $Q_{w,\,calc}$ (between 1 and 2).

Figure 5 shows a similar comparison as in Figure 4, but for other averaged measurement data: $\dot{V}_w = 1,273.37$ litres/h, $\overline{T}'_{am} = 14.28$ °C and $\overline{T}'_w = 60.51$ °C. Due to the higher flow rate of water flow through a car radiator equal to $\dot{V}_w = 1273.37$ liters/h, the flow regime of water in pipes was turbulent. The water-side Reynolds number $Re_{w,u}$ in the first pass pipes was between 5238 and 5440. In the lower pass, the Reynolds number $Re_{w,l}$ was higher due to the smaller number of pipes and varied between 5820 and 5883.

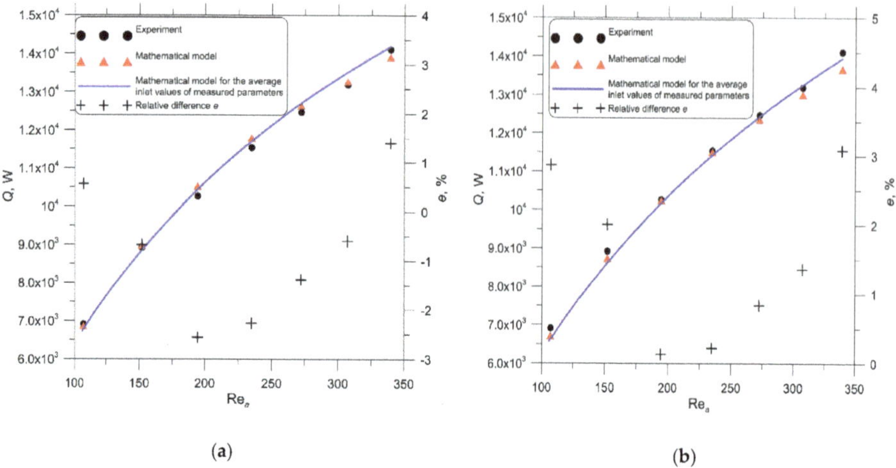

(a)   (b)

Figure 5. The thermal output Q of the car radiator; 1: $Q_{w,\,exp}$ values determined using the measurement results, 2: thermal output $Q_{w,\,calc}$ calculated with the air-side Nusselt number Nu_a obtained by CFD simulation (**a**) or empirical Equation (**b**), 3: thermal output $Q_{w,\,calc}$ obtained using the CFD-based correlation (**a**) or empirical correlation (**b**) and the average input data from seven measurement series: $\overline{\dot{V}}_w = 1273.37$ litres/h, $\overline{T}'_{am} = 14.28\ ^\circ C$ and $\overline{T}'_w = 60.51\ ^\circ C$, 4: the relative difference between $Q_{w,\,exp}$ and $Q_{w,\,calc}$.

The analysis of the results presented in Figure 5a,b shows that the thermal output of the car radiator calculated using a mathematical model is very close to the output determined experimentally, both for the relationship for the air-side HTC determined by computer simulation (Figure 5a) and for the empirical correlation for HTC determined by experimental tests (Figure 5b). The relative differences e between the measured and calculated capacities are in the range of about (-2.5%) to 4.5% (Figure 5).

It was found that the correlation for the air-side Nusselt number Nu_a determined for the entire double-row radiator based on CFD modelling provides a perfect match between the calculated and measured thermal output (Figures 4 and 5). Then, the heat flow transferred in the first and second rows of pipes in both radiator passes was calculated, taking into account different first and second-row HTCs. The following equations were applied to calculate the capacity in each row of pipes in both passes in the car radiators:

$$Q_{1,jHTC} = \frac{\dot{m}_w}{2}\,\bar{c}_w\left(T'_w - T''_{w,1}\right)\, j = 1,2 \tag{50}$$

$$Q_{2,jHTC} = \frac{\dot{m}_w}{2}\,\bar{c}_w\left(T'_w - T''_{w,2}\right)\, j = 1,2 \tag{51}$$

$$Q_{3,jHTC} = \frac{\dot{m}_w}{2}\,\bar{c}_w\left(T_{wm} - T''_{w,3}\right)\, j = 1,2 \tag{52}$$

$$Q_{4,jHTC} = \frac{\dot{m}_w}{2}\,\bar{c}_w\left(T_{wm} - T''_{w,4}\right)\, j = 1,2 \tag{53}$$

The total capacity of the car radiator was evaluated as follows:

$$Q_{t,jHTC} = \dot{m}_w\,\bar{c}_w\left(T'_w - T''_w\right)\, j = 1,2 \tag{54}$$

The designations $Q_{i,1HTC}\ i = 1,\ldots,4$ mean the thermal capacity of the specific pipe row (Figure 1) with a uniform HTC throughout the car radiator given by Equation (32). The symbol $Q_{i,2HTC}\ i = 1,\ldots,4$ stands for the thermal capacity in a particular row, with different HTCs in the first and second pipe rows. The first-row HTC was evaluated, applying Equation (30) and in the second row by using Equation (31). When the HTC on the air-side is constant across the entire radiator, then in Equations (50)–(54), $j = 1$ is assumed, and if the HTCs in the first and second tube rows are different, then $j = 2$ should be used. The relative differences e_i between the thermal capacity with different HTCs and the capacity with constant HTC across the entire car radiator were also determined:

$$e_i = \frac{Q_{i,2HTC} - Q_{i,1HTC}}{Q_{i,2HTC}} \cdot 100 \; i = 1,\ldots,4 \tag{55}$$

The relative difference e_t for the entire PFTHE was determined by a similar formula

$$e_t = \frac{Q_{t,2HTC} - Q_{t,1HTC}}{Q_{t,2HTC}} \cdot 100 \tag{56}$$

A comparison of the thermal output of the specific tube rows and the radiator heat output for the uniform HTC throughout the radiator and different HTCs in individual tube rows are depicted in Figure 6a for $\overline{\dot{V}}_w = 326.06$ liters/h and in Figure 6b for $\overline{\dot{V}}_w = 1273.37$ litres/h.

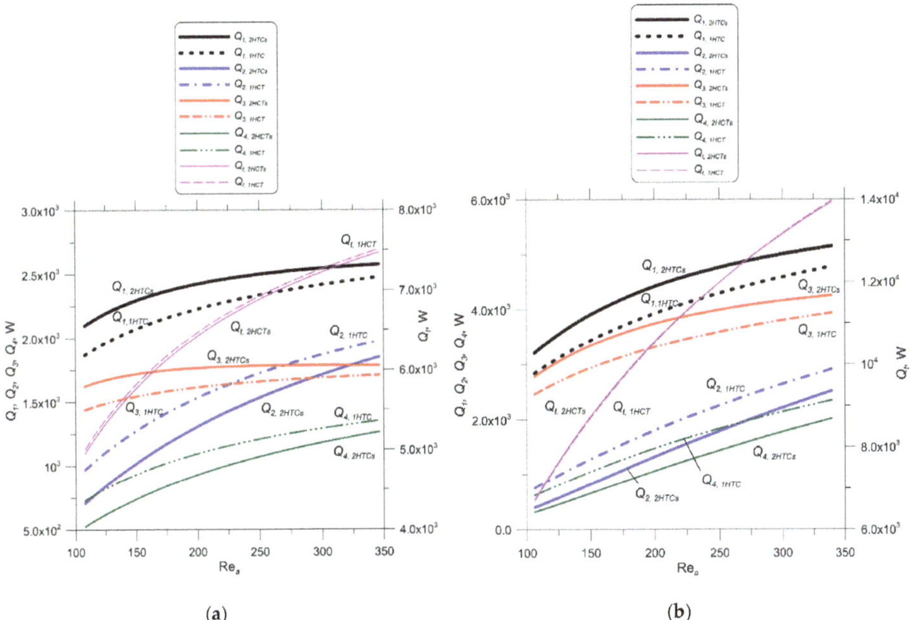

Figure 6. Comparison of thermal outputs of the specific pipe rows and whole-car radiator considering that the air-side HTC is constant in both pipe rows with the corresponding thermal outputs for different HTCs in the first and second pipe rows; (a) $\dot{V}_w = 326.06$ litres/h, $\overline{T}'_{am} = 13.62\,°C$ and $\overline{T}'_w = 59.61\,°C$, (b) $\dot{V}_w = 1273.37$ litres/h, $\overline{T}'_{am} = 14.28\,°C$ and $\overline{T}'_w = 60.51\,°C$; $Q_{i,1HTC}, i = 1,\ldots,4$: thermal output of the specific pipe row (Figure 1) adopting the same correlation (32) for the Nusselt number on the air side, $Q_{i,2HTC}, i = 1,\ldots,4$: thermal outputof a specific pipe row for different correlations for the air side Nusselt number (Figure 1); Equation (30) was used for the first row and Equation (31) for the second row of tubes, $Q_{t,1HTC}$: car radiator output for constant HTC calculated from Equation (32) for all rows of pipes, $Q_{t,2HTC}$: car radiator output for different HTCs; Equation (30) was used for the first tube row and Equation (31) for the second tube row.

The inspection of the results presented in Figure 6a,b shows that the thermal output of the radiator is almost identical for a uniform HTC throughout the radiator and different HTCs in both rows of pipes. As shown in Figures 5 and 6, the radiator capacity determined by applying an analytical model of the engine cooler using the CFD-based air-side equation for the Nusselt number is in line with the measurements. It can also be seen that the thermal output of the first and second tube rows in both passes of the radiator is significantly higher for different correlations for the air-side Nusselt compared to the heat flow rates obtained assuming the same HTC in both pipe rows.

The heat capacity of the heat exchanger is almost identical to the same and different HTCs for both rows of pipes (Figures 5 and 6). Uniform HTC for two rows of pipes was found from the condition of equal air temperature rise in the whole exchanger determined from CFD modelling and analytical formula [19]. Considering that the air temperature gain in both pipe rows is equal to the sum of the temperature rise in the first and second pipe rows, it can be expected that the heat output of the entire heat exchanger will be almost identical to the uniform and different HTCs. Figure 7a,b illustrate the higher heat absorption by the air in the first pipe row and the considered reduction of the heat flow rate in the second pipe row if different correlations for the Nusselt numbers are adopted compared to the respective outputs of the individual rows with uniform HTC.

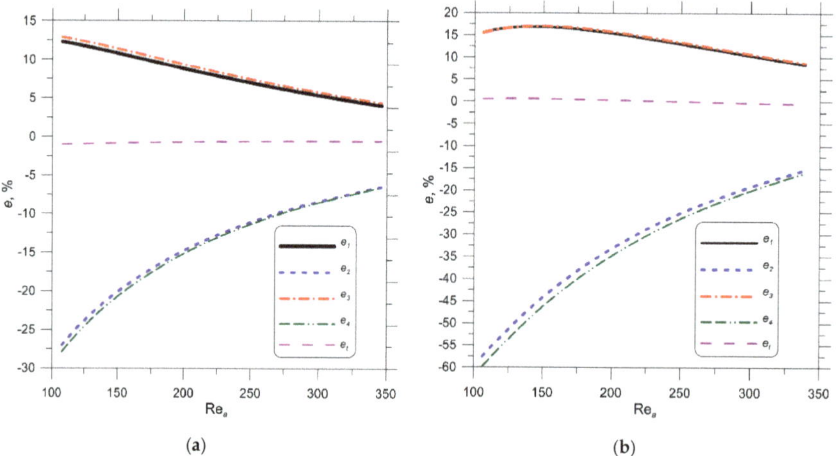

(a)   (b)

Figure 7. The relative differences between thermal outputs of specific tube rows for different and constant HTCs; (**a**) $\dot{\overline{V}}_w = 326.06$ litres/h, $\overline{T}'_{am} = 13.62\ °C$ and $\overline{T}'_w = 59.61\ °C$, (**b**) $\dot{\overline{V}}_w = 1273.37$ litres/h, $\overline{T}'_{am} = 14.28\ °C$ and $\overline{T}'_w = 60.51\ °C$.

Figure 7a,b show that considerable discrepancies in the exchanged heat flow through the first and second rows of pipes for uniform and different HTCs are higher for a larger volume flow rate passing through the cooler. The differences in the thermal capacities of individual pipe rows with equal and different HTCs for both pipe rows become smaller as the Reynolds number on the air side increases.

7.2. Engine Cooler Made of Circular Tubes

Figure 8 depicts a comparison of the thermal output of the car radiator for the CFD-based air-side correlation (35) and empirical correlation (45) for averaged measurement data: $\overline{w}_0 = 2.27$ m/s, $\overline{T}'_{am} = 8.24\ °C$, and $\overline{T}'_w = 70.56\ °C$. The volume flow rate of water $\dot{V}_w$ at the radiator inlet varied from 309 to 2,406 litres/h. The Reynolds number Re_a on the air side changed from 482.5 to 489.5, and the water-side Reynolds number $Re_{w,u}$ in the first (upper) pass of the cooler varied from 1,834 to 16,084. Examination of the results reported in Figure 8 reveals that the capacity of a round-tube radiator calculated for the empirical correlation with the air-side Nusselt number exceeds the capacity of the radiator calculated using the correlation with the Nusselt number on the air side obtained by CFD modelling. The relative difference e calculated using Equation (49) is about 5% for a Reynolds number $Re_{w,u}$ on the water side of 1800 and about 10% for a Reynolds number of 16,000 (Figure 8).

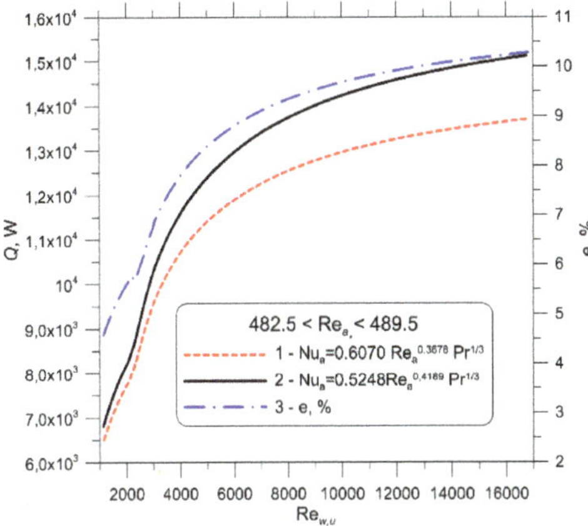

Figure 8. The heat flow rate $Q_{t,1HTC}$ from water to air in an engine cooler constructed from round tubes with an equal HTC on the air side of the entire radiator versus the first-pass Reynolds number $Re_{w,u}$ on the water side; 1: air-side correlation for Nu_a based on CFD modelling, 2: air-side relationship for Nu_a based on experimental data, 3: relative difference e.

The analysis of the results shown in Figure 9a indicates that the relative difference e between the radiator output determined experimentally and that calculated using a mathematical model of the radiator with a correlation to the air-side Nusselt number obtained by CFD modelling is between 3% and about 8.3%. In the case when the air-side relationship is determined experimentally, the relative difference e is smaller and is in the range from −4% to 0% (Figure 9b). Figure 10 presents a comparison of the thermal outputs of the individual pipe rows assuming the same air-side HTC in the analytical model of the radiator and different HTCs in the first and second pipe rows.

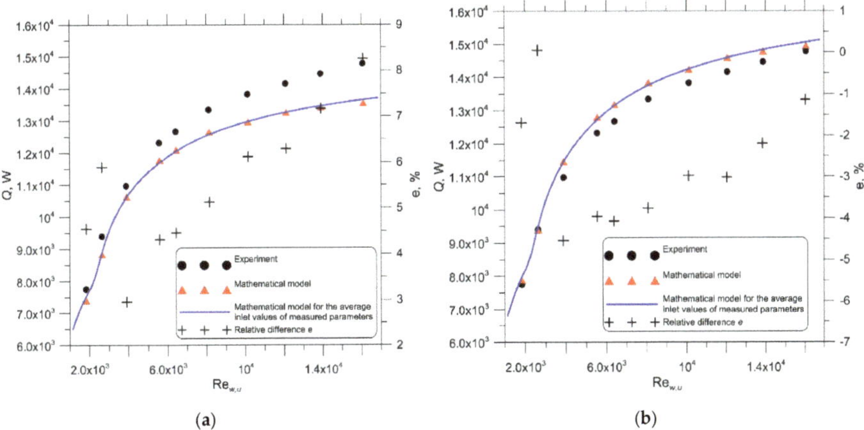

Figure 9. The comparison of radiator output Q versus the water-side Reynolds number $Re_{w,u}$ obtained experimentally and using the mathematical model of the engine cooler with the uniform air-side Nusselt number Nu_a determined by (**a**) CFD modelling and (**b**) empirical correlation; e: relative difference.

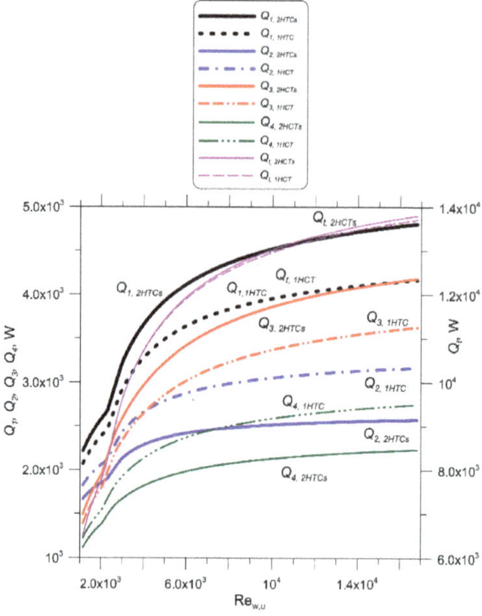

Figure 10. Comparison of thermal capacities of specific pipe rows and the whole-car radiator supposing that HTC on the air-side is uniform in both pipe rows with the corresponding thermal capacities obtained for different HTCs in the first and second pipe rows; $\overline{w}_0 = 2.27$ m/s, $\overline{T}'_{am} = 8.24\ ^\circ$C, and $\overline{T}'_w = 70.56\ ^\circ$C; $Q_{i,1HTC}$, $i = 1,\ldots,4$: thermal output of the specific pipe rows (Figure 1) supposing the same Equation (32) for the Nusselt number on the air-side throughout the radiator, $Q_{i,2HTC}$, $i = 1,\ldots,4$: thermal outputs of individual pipe rows (Figure 1) for different equations for the air-side Nusselt in both pipe rows; Equation (30) was used for the first row of tubes and Equation (31) for the second row, $Q_{t,1HTC}$: car radiator output calculated using Equation (32) for all rows of pipes, $Q_{t,2HTC}$: car radiator output calculated using different heat transfer correlations for the first and second tube rows.

The air in the first pipe row has a much larger heat flow rate assuming different HTCs in the first and second pipe rows compared to the heat flow rate determined considering an even HTC throughout the heat exchanger (Figures 10 and 11).

The relative differences e_1 and e_3 are approximately 7.5% for $\mathrm{Re}_{w,u}$ equal to 1000 and increase with $\mathrm{Re}_{w,u}$ to about 15% for the Reynolds number $\mathrm{Re}_{w,u}$ equal to 18,000. The relative differences of e_2 and e_4 are approximately -8% for $\mathrm{Re}_{w,u}$ equal to 1000 and decrease with $\mathrm{Re}_{w,u}$ to approximately -18% for the Reynolds number $\mathrm{Re}_{w,u}$ equal to 18,000. The results depicted in Figures 10 and 11 show that the radiator thermal output is almost identical for the same and different HTCs on both rows of pipes, despite the various capacities of the first and second rows of tubes.

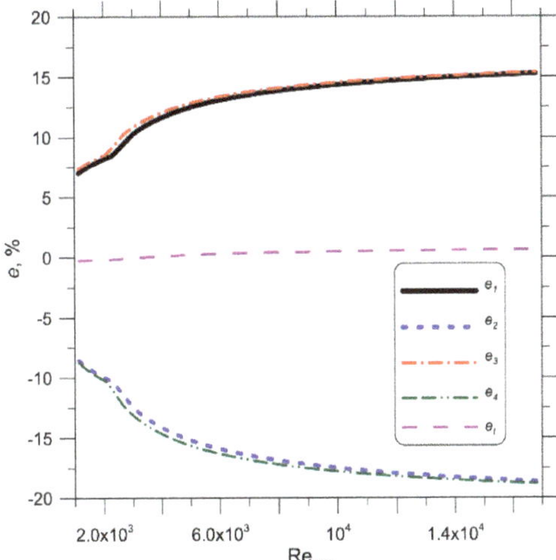

Figure 11. The relative differences e_i and e_t between thermal outputs of individual tube rows and the entire car radiator for different and uniform HTCs calculated using Equations (55) and (56); $\overline{w}_0 = 2.27$ m/s, $\overline{T}'_{am} = 8.24\,°C$, and $\overline{T}'_w = 70.56\,°C$.

8. Conclusions

An analytical model of a two-pass automobile radiator with two rows of pipes has been developed. Correlations determined by CFD modelling should be used to determine the air-side HTC. Radiators made of round and oval pipes were analysed. Mathematical models of car radiators were proposed, in which only one average HTC for the entire heat exchanger was used, as well as various correlations for calculating the air-side HTCs for the specific of rows of pipes. Calculation of the PFTHE with various HTCs for each row of tubes makes it possible to calculate more accurately the thermal output of individual rows of pipes, which, for example, allows designing a PFTHE with an optimum tube row number. The relationships based on the numerical solution of the equations of momentum and energy conservation for the fluid were used to calculate the HTC on the inner surface of the pipes. The water flow in the tubes may be laminar, transitional, or turbulent. The results of the calculations of both tested car radiators were compared with experimental data. The agreement between calculation and measurement results is excellent. The proposed method of calculating PFTHEs is attractive because of the elimination of costly experimental research necessary for the experimental determining of the heat transfer correlations on the air and water side.

The main achievements presented in the article are as follows:

- An exact analytical model of a single-pass double-row heat exchanger with different heat transfer coefficients on the first and second rows of pipes.
- Exact analytical model of a two-row, two-pass car radiator (plate-fin and tube heat exchanger (PFTHE)) with different heat transfer coefficients for the first and second tube rows.
- Calculation method of PFTHE without using empirical heat transfer correlations on the water and air sides.
- The results of extensive experimental research on two car radiators, one of which is made of oval tubes and the other of round tubes.
- The water flow in the pipes can be laminar, transient or turbulent while maintaining the continuity of the heat transfer coefficient when changing the flow regime. CFD modeling of all three flow ranges would be difficult.

Author Contributions: Conceptualization, J.T. and D.T.; methodology, D.T.; software, M.T. and D.T.; validation, J.T., D.T. and M.T.; formal analysis, J.T.; investigation, M.T. and D.T.; resources, M.T.; data curation, M.T.; writing—original draft preparation, D.T. and J.T.; writing—review and editing, J.T., D.T. and M.T.; visualization, D.T. and M.T.; supervision, J.T.; project administration, D.T.; All authors have read and agreed to the published version of the manuscript.

Funding: This research received no external funding.

Conflicts of Interest: The authors declare no conflict of interest.

References

1. Shah, R.K.; Sekuli, D.P. *Fundamentals of Heat Exchanger Design*; John Wiley & Sons: Hoboken, NJ, USA, 2003.
2. Kakaç, S.; Liu, H.; Pramuanjaroenkij, A. *Heat Exchangers. Selection, Rating, and Thermal Design*, 3rd ed.; CRC Press-Taylor and Francis Group: Boca Raton, FL, USA, 2012.
3. Kuppan, T. *Heat Exchanger Design Handbook*, 2nd ed.; CRC Press Taylor and Francis Group: Boca Raton, FL, USA, 2013.
4. Taler, D. *Numerical Modelling and Experimental Testing of Heat Exchangers*; Springer: Berlin/Heidelberg, Germany, 2019.
5. Kim, N.-H.; Yun, J.-H.; Webb, R.L. Heat Transfer and Friction Correlations for Wavy Plate Fin-and-Tube Heat Exchangers. *J. Heat Transf.* **1997**, *119*, 560–567. [CrossRef]
6. Kim, N.H.; Youn, B.; Webb, R.L. Air-Side Heat Transfer and Friction Correlations for Plain Fin-and-Tube Heat Exchangers With Staggered Tube Arrangements. *J. Heat Transf.* **1999**, *121*, 662–667. [CrossRef]
7. Wang, C.-C.; Hsieh, Y.-C.; Lin, Y.-T. Performance of Plate Finned Tube Heat Exchangers Under Dehumidifying Conditions. *J. Heat Transf.* **1997**, *119*, 109–117. [CrossRef]
8. Halıcı, F.; Taymaz, I.; Gunduz, M.; Halici, F. The effect of the number of tube rows on heat, mass and momentum transfer in flat-plate finned tube heat exchangers. *Energy* **2001**, *26*, 963–972. [CrossRef]
9. Taler, D. Mathematical modeling and experimental study of heat transfer in a low-duty air-cooled heat exchanger. *Energy Convers. Manag.* **2018**, *159*, 232–243. [CrossRef]
10. Taler, D.; Taler, J. Prediction of heat transfer correlations in a low-loaded plate- fin-and-tube heat exchanger based on flow-thermal tests. *Appl. Therm. Eng.* **2019**, *148*, 641–649. [CrossRef]
11. Rich, D.G. The effect of the number of tube rows on heat transfer performance of smooth plate fin-and-tube heat exchangers. *ASHRAE Trans.* **1975**, *81 Pt 1*, 307–317. [CrossRef]
12. Marković, S.; Jaćimović, B.; Genić, S.; Mihailović, M.; Milovancevic, U.; Otović, M. Air side pressure drop in plate finned tube heat exchangers. *Int. J. Refrig.* **2019**, *99*, 24–29. [CrossRef]
13. McQuiston, F.C.; Parker, J.D.; Spitler, J.D. *Heating, Ventilating, and Air Conditioning Analysis and Design*, 6th ed.; Wiley: Hoboken, NJ, USA, 2005.
14. Webb, R.L.; Kim, N.H. *Principles of Enhanced Heat Transfer*, 2nd ed.; CRC Press: Boca Raton, FL, USA, 2005.
15. Sun, C.; Lewpiriyawong, N.; Khoo, K.L.; Zeng, S.; Lee, P.-S. Thermal enhancement of fin and tube heat exchanger with guiding channels and topology optimisation. *Int. J. Heat Mass Transf.* **2018**, *127*, 1001–1013. [CrossRef]
16. Li, M.; Zhang, H.; Zhang, J.; Mu, Y.; Tian, E.; Dan, D.; Zhang, X.; Tao, W. Experimental and numerical study and comparison of performance for wavy fin and a plain fin with radiantly arranged winglets around each tube in fin-and-tube heat exchangers. *Appl. Therm. Eng.* **2018**, *133*, 298–307. [CrossRef]
17. Nagaosa, R. Turbulence model-free approach for predictions of air flow dynamics and heat transfer in a fin-and-tube exchanger. *Energy Convers. Manag.* **2017**, *142*, 414–425. [CrossRef]
18. Kearney, S.P.; Jacobi, A. Local Convective Behavior and Fin Efficiency in Shallow Banks of In-Line and Staggered, Annularly Finned Tubes. *J. Heat Transf.* **1996**, *118*, 317–326. [CrossRef]
19. Taler, D.; Taler, J.; Trojan, M. Thermal calculations of plate-fin-and-tube heat exchangers with different heat transfer coefficients on each tube row. *Energy* **2020**, *203*, 117806. [CrossRef]
20. Gnielinski, V. Heat transfer in pipe flow, chapter G1. In *VDI Heat Atlas*, 2nd ed.; Springer: Berlin/Heidelberg, Germany, 2010; pp. 691–700.

21. Taler, D. A new heat transfer correlation for transition and turbulent fluid flow in tubes. *Int. J. Therm. Sci.* **2016**, *108*, 108–122. [CrossRef]
22. Taler, D. Determining velocity and friction factor for turbulent flow in smooth tubes. *Int. J. Therm. Sci.* **2016**, *105*, 109–122. [CrossRef]

MDPI
St. Alban-Anlage 66
4052 Basel
Switzerland
Tel. +41 61 683 77 34
Fax +41 61 302 89 18
www.mdpi.com

Energies Editorial Office
E-mail: energies@mdpi.com
www.mdpi.com/journal/energies